STP 1130

Rapid Load Fracture Testing

Ravinder Chona and William R. Corwin, editors

ASTM Publication Code Number (PCN)
04-011300-30

ASTM
1916 Race Street
Philadelphia, PA 19103

Library of Congress Cataloging-in-Publication Data

Rapid load fracture testing/Ravinder Chona and William R. Corwin, editors.
(ASTM STP; 1130)
"ASTM publication code number (PCN) 04-011300-30."
Includes bibliographical references and index.
ISBN 0-8031-1429-X
1. Steel—Testing. 2. Metals—Impact testing. 3. Steel—Fracture. I. Chona, Ravinder.
II. Corwin, W. R. III. Series: ASTM special technical publication; 1130.
TA465.R37 1992
620.1′76—dc20

91-45387
CIP

Photocopy Rights

Peer Review Policy

Each paper published in this volume was evaluated by three peer reviewers. The authors addressed all of the reviewers' comments to the satisfaction of both the technical editor(s) and the ASTM Committee on Publications.

The quality of the papers in this publication reflects not only the obvious efforts of the authors and the technical editor(s), but also the work of these peer reviewers. The ASTM Committee on Publications acknowledges with appreciation their dedication and contribution to time and effort on behalf of ASTM.

Printed in Baltimore, MD
March 1992

Foreword

The symposium on Rapid Load Fracture Testing was presented in San Francisco, California, on 23 April 1990. ASTM Committee E-24 on Fracture Testing sponsored the symposium. Ravinder Chona, Texas A&M Univeristy, and William R. Corwin, Oak Ridge National Laboratory, served as chairmen of the symposium and editors of the resulting publication.

Contents

Overview

The Symposium on Rapid Load Fracture Testing was organized by ASTM Task Group E-24.01.06 on Dynamic Fracture Toughness and Crack Arrest and was held in April 1990 in conjunction with the semiannual standards development meetings of ASTM Committee E 24 on Fracture Testing. The aim of the symposium was to review the state of the art with regard to the use of rapid loading to determine the fracture toughness behavior of ferritic steels in the ductile-to-brittle transition region. In particular, the symposium focused on test methods that could: reduce the amount of data scatter; illustrate or establish any relationships between K_{Ic}, K_{Id}, and/or K_{Ia}; provide lower-bound measures of fracture toughness; and improve the efficiency of testing with material of limited availability.

The papers presented at the symposium, and published in this volume following the usual ASTM peer-review process, described a variety of test techniques, specimen geometries, and data acquisition, analysis, and interpretation methods, all generally suited to loading times to failure of the order of 1 to 2 milliseconds or less. This may, at first, be somewhat puzzling to the reader, since it is generally recognized that the structural applications of interest would be unlikely to involve loadings at comparable rates. The rationale is, however, as follows. It has been demonstrated that, within the ductile-to-brittle transition region, the crack arrest fracture toughness, K_{Ia}, for a given temperature, is consistently below the initiation toughness, K_{Ic}, of the material, and can potentially serve as a conservative, lower-bound estimate of K_{Ic}. It has also been demonstrated that, at temperatures close to and below the nil ductility temperature, NDT, the values of K_{Ic} obtained from tests conducted with rapid loading times, following Annex A-7 of the ASTM Test for Plane-Strain Fracture Toughness of Metallic Materials (E 399) provide close estimates of K_{Ia}, with the required loading time being of the order of 5 milliseconds at temperatures close to the NDT. The usefulness of rapid loading in transition region testing, therefore, lies more in the increased probability for initiating a rapid, unstable, cleavage-type fracture, with little or no prior stable crack extension, when performing material characterization tests with small, laboratory-sized specimens. A brief summary of the contents of this volume follows.

A major area of interest from an applications standpoint is the establishment of safe operating pressure-temperature relationships for nuclear reactor pressure vessels. The paper by Server and Mager, which leads off this volume, provides an overall perspective of how the information obtained from this type of testing might be used and summarizes the current thinking regarding operating regulations from the viewpoint of the nuclear industry.

The next group of seven papers discusses a variety of loading techniques and specimen geometries as well as various methods for interpreting dynamically recorded signals to obtain fracture parameters. The first subgroup of three papers, by Irwin et al., Couque et al., and Homma et al., describe three rather different techniques for achieving cleavage fracture using short duration stress wave loading, while the second subgroup of four papers, by Kirk et al., KarisAllen and Morrison, Böhme, and McConnell, all address various aspects of testing using impact-loaded bend bars.

A somewhat different topic is addressed in the next paper by Tregoning et al., which describes an optical technique for monitoring the CTOD before and following initiation of a dynamically loaded, stationary crack.

The next two papers both use the ASTM Test for Determining the Plane-Strain Crack Arrest Fracture Toughness K_{Ia} of Ferritic Steels (E 1221): Varga and Schneeweiss describe crack-arrest toughness measurements using instrumented Charpy V-notch specimens and compare their results to those obtained with standard K_{Ia} specimens, while Underwood et al., discuss the application of ASTM Test E 1221 to a ship steel and compare the results for K_{Ia} to the values of K_{Ic} for the same material.

The final paper, by McGillivray and Cannon, describes a test method under development in the United Kingdom for determining the dynamic fracture toughness of metallic materials at loading rates that can be achieved using an impact-loading arrangement.

The overall goal of the symposium was to bring together a group of active researchers addressing the various aspects of using rapid-loading techniques when performing fracture toughness evaluations and to see if the presentations and subsequent discussions would indicate that a standardization effort was warranted at the present time. Considerable interest in the topic was evident, but more time is clearly needed before a consensus can be established on the most suitable methods for standardization activities. The potential usefulness of rapid loading for achieving the goal of reliable, lower-bound, transition region fracture toughness measurements is felt to be well documented by the contents of this volume, and it is hoped that this collection of papers will be the first in an ongoing series that will benchmark progress towards a useful and necessary standard.

Ravinder Chona
Texas A&M University, College Station, TX; symposium cochairman and coeditor.

William R. Corwin
Oak Ridge National Laboratory, Oak Ridge, TN; symposium cochairman and coeditor.

William L. Server[1] *and Thomas R. Mager*[2]

Irradiated Dynamic and Arrest Fracture Toughness Compared to Lower-Bound Predictions

REFERENCE: Server, W. L. and Mager, T. R., **"Irradiated Dynamic and Arrest Fracture Toughness Compared to Lower-Bound Predictions,"** *Rapid Load Fracture Testing, ASTM STP 1130,* Ravinder Chona and William R. Corwin, Eds., American Society for Testing and Materials, Philadelphia, 1992, pp. 1–8.

ABSTRACT: Pressure-temperature operating curves for nuclear reactor pressure vessels are based upon a lower-bound fracture toughness curve which bounds rapid load dynamic initiation and crack arrest fracture toughness data. The ASME Boiler and Pressure Vessel Code defines the reference toughness (K_{IR}) curve as this lower bound, and this K_{IR} curve was developed solely from unirradiated dynamic and arrest fracture toughness data from one heat of SA533B-1 steel (HSST Plate 02) and two heats of SA508-2 steel. The effects of radiation embrittlement on the shape and shift of the K_{IR} curve to account for the increase in reference temperature is thought to be conservative, but this conservatism has not been fully verified. This study reviews available data from past dynamic and arrest toughness tests on irradiated vessel steels from test reactor irradiations and compares the data to the shifted K_{IR} curve using the transition temperature shift approach detailed in Regulatory Guide 1.99, Revision 2. Dynamic initiation and crack arrest fracture toughness data are available from only a few irradiated large specimen tests (that is, test specimens with thicknesses greater than about 51 mm [2 in.]); small specimen tests (including precracked Charpy) are used for the other comparisons. The limited results indicate that the Regulatory approach for shifting the K_{IR} curve is very conservative even when the Regulatory Guide 1.99, Revision 2 "margin term" is not used and a correction for fluence rate is ignored. No change in shape for the dynamic toughness and arrest data (in particular for low upper shelf materials) was observed.

KEYWORDS: embrittlement, pressure vessel steel, fracture toughness, dynamic toughness, crack arrest, transition temperature, radiation damage

The American Society of Mechanical Engineers (ASME) has published in the Boiler and Pressure Vessel Code, Appendix G to Section III, a procedure for obtaining the allowable loading in pressure-temperature space for ferritic pressure-retaining materials of Class 1 components, such as the reactor pressure vessel. The specified procedure in the ASME Code, Section III, Appendix G is based upon the principles of linear elastic fracture mechanics. Section III, Appendix G presents a reference stress intensity factor (K_{IR}) as a function of temperature based upon the lower bound of static initiation (K_{Ic}), dynamic initiation (K_{Id}), and crack arrest (K_{Ia}) fracture toughness values. Appendix G also specifies a postulated defect of one-quarter thickness to be used in determining the allowable loading and defines methods which can be used to calculate the applied stress intensity factor.

The K_{IR} curve additionally is included in the ASME Code, Appendix G to Section XI (Inservice Inspection) to cover pressure-temperature curves after initial plant operation. The flaw

[1] Vice president, ATI Consulting, 2010 Crow Canyon Place—Suite 160, San Ramon, CA 94583.

[2] Consulting engineer, Westinghouse Electric Corp., Nuclear and Advanced Technology Division, P.O. Box 2728, Pittsburgh, PA 15230.

evaluation procedures contained in Appendix A of Section XI also use the K_{IR} curve, but it is referred to as the K_{Ia} curve for the assessment of discovered defects which are larger than the Section XI acceptance standards. All of these Section III and XI Appendices to the ASME Code are nonmandatory except when implementation requirements are specified by the Nuclear Regulatory Commission (NRC) through the Code of Federal Regulations (10 CFR Part 50), Appendix G. Currently, Appendix G to Section III is mandatory, and NRC is in the process of making Appendix G to Section XI mandatory. Appendix A is not mandatory, although an analysis of the type specified in Appendix A is required for assessing significant discovered defects.

The use of the K_{IR} curve involves a reference nil-ductility transition temperature (RT_{NDT}) which indexes the K_{IR} curve to the temperature scale. The value RT_{NDT} is defined as the greater of the nil-ductility transition temperatures (NDTT per the ASTM Test for Conducting Drop-Weight Test to Determine Nil-Ductility Transition Temperature of Ferritic Steels [E 208]), and the temperature 33°C (60°F) less than a lower-bound 68-J (50-ft-lbf) energy/0.89-mm (35-mils) lateral expansion temperature as determined using Charpy V-notch test specimens oriented in the transverse direction (normal to the rolling or major working direction of the material). The actual determination of RT_{NDT} for the unirradiated condition is specified in Section III of the ASME Code, NB-2300.

Typically, the Charpy energy and lateral expansion lower bounds are developed by testing three Charpy specimens at NDTT + 33°C (NDTT + 60°F) and assuring that no energy or lateral expansion value is below 68 J (50 ft-lbf) or 0.89 mm (35 mils), respectively; if one or more values fall below these levels, a series of three Charpy specimens are tested at increasing increments of 5.6°C (10°F) until the requirements are met. Since neutron irradiation increases the nil-ductility transition temperature and reduces the fracture toughness of ferritic materials, assurance of safety margins must be maintained by adjusting the lower-bound K_{IR} curve in accordance with the degree of embrittlement. The procedure typically used is to adjust the value of RT_{NDT} by adding an increment which represents the shift in measured Charpy V-notch transition temperature at the 41-J (30-ft-lbf) level. This shift in the K_{IR} curve to account for radiation embrittlement is *not* based upon measured fracture toughness data from commercial surveillance programs, but is generally based upon a conservative estimate of the Charpy shift or a combination of measured Charpy shift from surveillance results and an added Regulatory margin term. Additionally, the shape of the K_{IR} curve is assumed to be constant after irradiation.

The purpose of this paper is to review briefly the original K_{IR} curve data and address the issues associated with post-irradiation fracture toughness data. In particular, the degree of conservatism in the shifted K_{IR} curve approach will be assessed. Only dynamic initiation and crack arrest data will be considered in this review.

K_{IR} Reference Stress Intensity Factor

The lower-bound curve developed by the Pressure Vessel Research Committee (PVRC) of the Welding Research Council [*1*] (which was subsequently incorporated into the ASME Code, Section III, Appendix G and Section XI, Appendices A and G) was expressed in the form:

$$K_{IR} = 26.777 + 1.223 \exp [0.0145 (T - RT_{NDT} + 160)] \tag{1}$$

where the test temperature T and RT_{NDT} are in degrees Fahrenheit and K_{IR} is in units of ksi-in$^{1/2}$. When converted to metric units, Eq 1 becomes:

$$K_{IR} = 29.425 + 1.343 \exp [0.0261 (T - RT_{NDT} + 88.89)] \tag{2}$$

where T and RT_{NDT} are now in degrees Celsius and K_{IR} is in units of MPa-m$^{1/2}$. The K_{IR} curve is

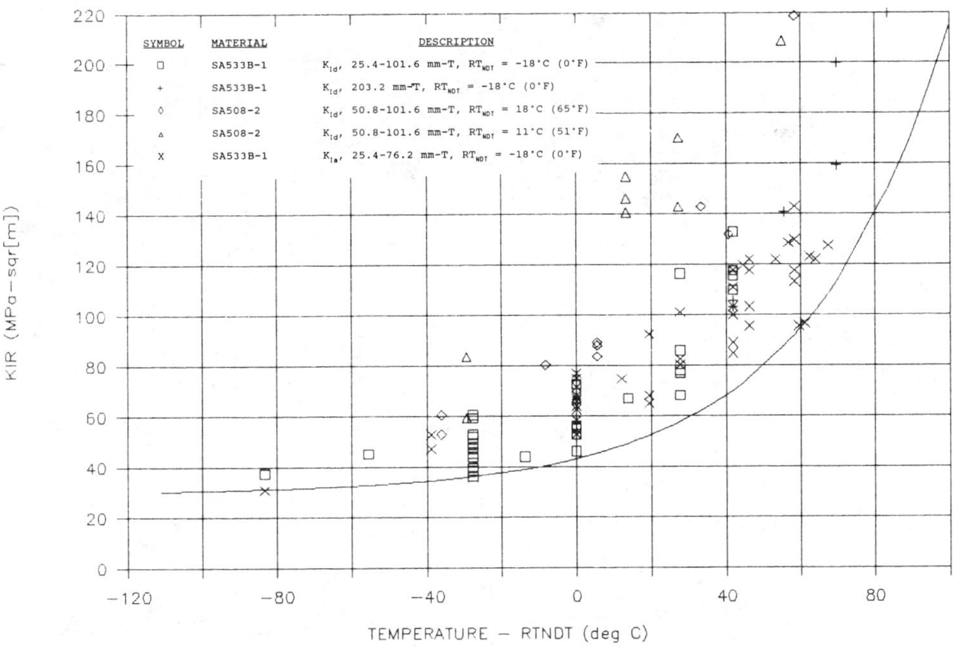

FIG. 1—*Lower-bound* K_{IR} *curve developed from dynamic initiation and crack arrest fracture toughness data.*

shown in Fig. 1, and the dynamic and crack arrest fracture toughness data [*1,2*] used to derive this lower-bound curve are also presented in this figure. Three heats of material were tested to form the basis for the K_{IR} curve: a heat of SA533B-1 steel (HSST Plate 02) and two heats of SA508-2 forging steel. The dynamic fracture toughness data for the three heats were generated by Westinghouse [*1–3*], and the legend for Fig. 1 provides specimen size and material differentiation information.

Thickness of test specimens is denoted using the metric measure and the letter T; that is, 25.4 mm-T means 25.4 mm (1.00 in.) thickness. The K_{Ia} data were generated only for the SA533B-1 heat, HSST Plate 02 [*4*]. The initial unirradiated RT_{NDT} for HSST Plate 02 was determined to be −18°C (0°F) using the ASME Code procedure; later investigations for this heat of material [*5*] suggest that the RT_{NDT} could be as high as 4°C (40°F) which would shift the data 22°C (40°F) to the left making the K_{IR} curve conservative compared to the position of the adjusted data which were used to derive the original lower bound. As can be seen in Fig. 1, the highest temperature portion of the lower bound is established from the lowest points at $[T - RT_{NDT}] = 61°C$ (110°F), which are crack arrest toughness measurements for HSST Plate 02. The lowest temperature portion of the bound is established from both crack arrest and dynamic initiation toughness measurements for the SA533B-1 steel (HSST Plate 02). Although not shown on this figure, valid static fracture toughness data [*6*] were generated from small to thick section compact fracture specimens, and these data all fall significantly higher than the dynamic results.

Irradiated Fracture Toughness Data

The true test of using the K_{IR} curve after exposure of pressure vessel steels to high energy neutrons is comparing irradiated dynamic initiation and crack arrest fracture toughness data to

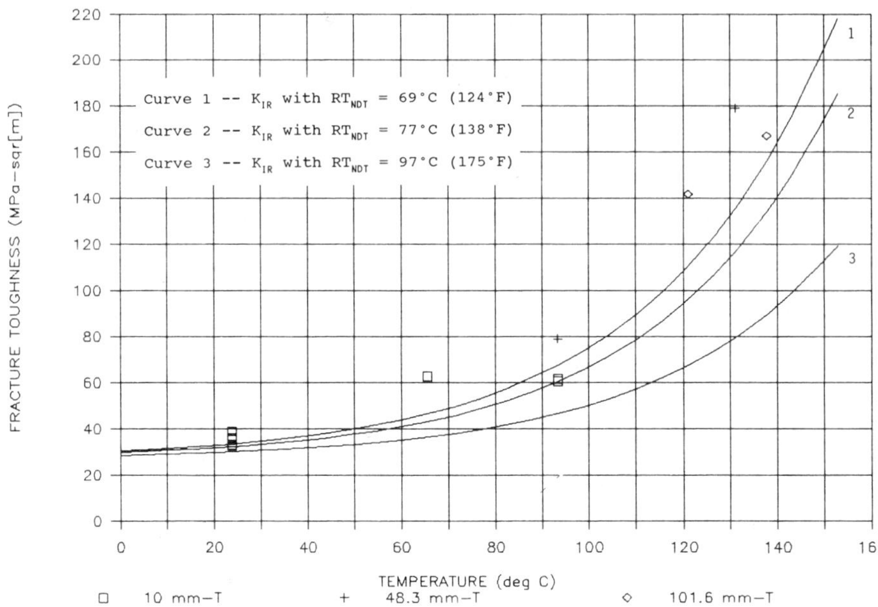

FIG. 2—*Unirradiated and irradiated dynamic fracture toughness data for an SA533B-1 steel (HSST Plate 02) compared to shifted* K_{IR} *curves.*

the predicted shifted lower bound using Regulatory Guide 1.99, Revision 2 [*7*]. Unfortunately, this comparison is complicated by the fact that the Charpy energy shift at 41 J (30 ft-lbf) for the Regulatory Guide approach is based upon power reactor conditions, whereas the measured fracture toughness data are developed from test reactor irradiations at a higher fluence rate. An adjustment in the power reactor test data is possible to account for this fluence rate difference by using a higher fluence exponent [*8*] in the damage equation (that is, using a fluence function exponent of 1/2 for test reactor data, as previously used in Regulatory Guide 1.99, Revision 1, instead of 1/4 to 1/3 as used in Regulatory Guide 1.99, Revision 2). This fluence function form becomes significant in that power reactor data may show a greater shift than test reactor results at higher fluences (that is, above 1×10^{19} n/cm^2 [for energies > 1 MeV]) and smaller shifts at lower fluences.

Shown in Fig. 2 are irradiated dynamic fracture toughness results [*5*] for three compact fracture specimen thicknesses: 10, 48.3, and 101.6 mm (0.394, 1.9, and 4 in.). The fluences for the three specimen sizes varied from 2.5 to 4.4×10^{19} n/cm^2. The K_{IR} curves shown represent three different irradiated RT_{NDT}s: (1) Regulatory Guide 1.99, Revision 2 prediction of a shift of 69°C (124°F) for a fluence of 2.5×10^{19} n/cm^2; (2) Regulatory Guide 1.99, Revision 2 shift prediction of 77°C (138°F) for a fluence of 4.4×10^{19} n/cm^2; and (3) the measured Charpy V-notch 41-J (30-ft-lbf) shift of 97°C (175°F). The Regulatory Guide predictions were based upon the mean predicted shift without a fluence rate correction nor the additional "margin term" indicated in the Regulatory Guide. Note that the data are just encompassed by the lower-bound K_{IR} curves (1 and 2) for the range of fluences using the Regulatory Guide approach.

If a fluence rate correction was made, the two K_{IR} curves would be shifted conservatively to the right by an additional 18 and 40°C (33 and 72°F). This additional shift amount brings the measured Charpy shift of 97°C (175°F) closer to an adjusted fluence rate predicted shift of 87 to 117°C (158 to 210°F). Therefore, the shift in the K_{IR} curve would be conservative without

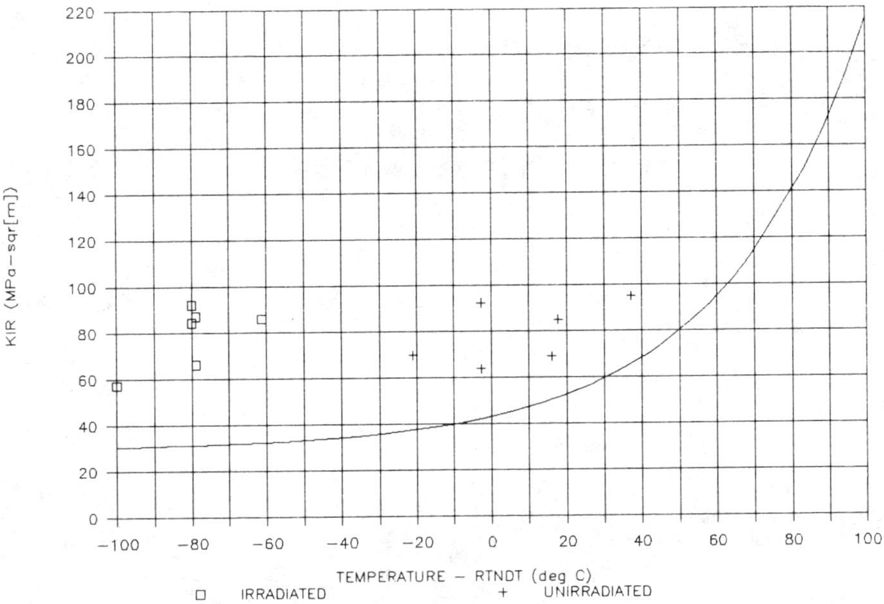

FIG. 3—*Comparison of unirradiated and irradiated crack arrest fracture toughness data for a Linde 80 weld metal with the* K_{IR} *curve.*

considering the effects of the added margin term in the Regulatory Guide method or the fact that these are test reactor results indicative of a stronger fluence function. Curve 3 illustrates significant conservatism as a lower bound for a shift equal to the measured Charpy 41-J (30-ft-lbf) shift of 97°C (175°F). These observations are consistent with the results of a previous study looking at the effects of irradiation on fracture toughness results as compared to the Regulatory Guide 1.99, Revision 2 shift to the ASME Code K_{IR} and static toughness (K_{Ic}) curves and to other reference toughness methods [*9*].

As indicated earlier, the K_{IR} curve is primarily fixed by the unirradiated crack arrest data from HSST Plate 02. Unfortunately, very few irradiated K_{Ia} tests have been performed. One study that was completed in the early 1980s [*10*] looked at two heats of SA533B-1 plate and two heats of submerged arc weld metal (Linde 0091 and Linde 80 flux types). The two plate materials differed primarily by the amount of copper in each plate, whereas the welds differed in the flux type and the amounts of copper and nickel. The higher copper material in each case was dubbed a "low upper shelf energy" steel, even though after irradiation (fluence = 1.4×10^{19} n/cm^2), the Charpy upper shelf energy for the weld was 87 and 125 J (64 and 92 ft-lbf) for the plate. These materials are not really what are considered low upper shelf energy materials since the upper shelf is significantly above 68 J (50 ft-lbf); the Code of Federal Regulations suggests that low upper shelf materials are those that have less than 102 J (75 ft-lbf) before irradiation and approach or go below 68 J (50 ft-lbf) after irradiation.

The data for the Linde 80 weld is the most complete and has the lowest irradiated upper shelf energy level. Figure 3 illustrates the Linde 80 weld K_{Ia} data for both the unirradiated and irradiated conditions [*10*] as compared to the K_{IR} curve. It is difficult to see any curve shape from the data as a result of the sparce data set and the limit of about 110 MPa $\cdot$ m$^{1/2}$ (100 ksi $\cdot$ in.$^{1/2}$) for valid data (because of the small test specimen size); thicknesses ranged from 25 to 51 mm (1 to 2 in.) for unirradiated and 16 to 29 mm (0.625 to 1.125 in.) for the irradiated specimens. The irra-

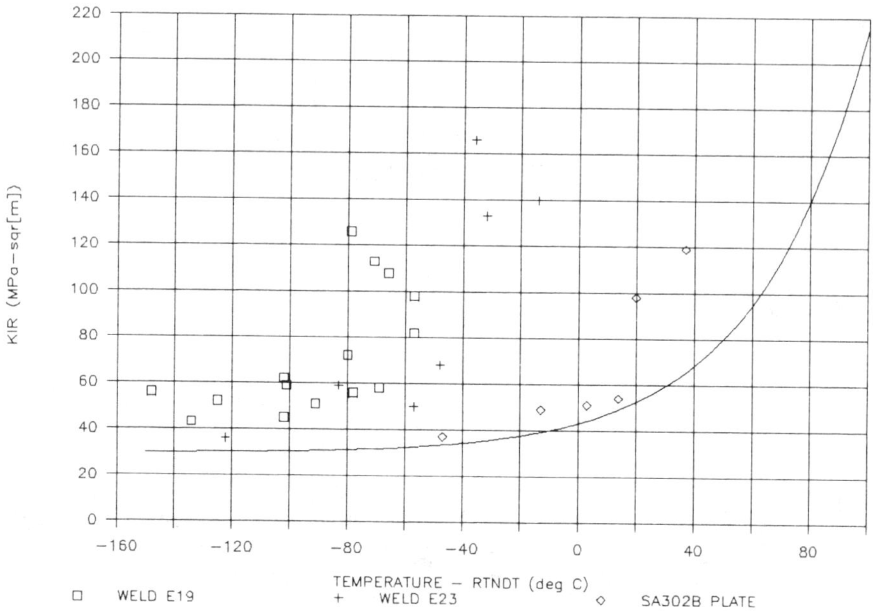

FIG. 4—*Comparison of dynamic initiation fracture toughness data for an SA302B steel and two Linde 80 welds with the* K_{IR} *curve.*

diated data are again shifted using Regulatory Guide 1.99, Revision 2 [*7*] without accounting for a fluence rate correction or inclusion of the "margin term." The data indicate no trend with temperature, but the data are conservatively on the upper and left side of the predicted lower bound K_{IR} curve.

A final comparison is made for two other Linde 80 welds and for a heat of SA302B steel (ASTM Correlation Monitor). In these cases, the only dynamic cleavage initiation data available for the irradiated condition is from precracked Charpy tests [*11*]. The main reason for selecting these three materials is due to their low upper shelf toughness after irradiation: the SA302B heat falls below 68 J (50 ft-lbf) in the transverse orientation after irradiation, and the other two heats of Linde 80 weld metal have upper shelf Charpy energies in the 81- to 95-J (60- to 70-ft-lbf) range. Figure 4 shows the irradiated precracked Charpy dynamic toughness data for these three materials compared to the K_{IR} curve shifted by Regulatory Guide 1.99 Revision 2, again without the fluence rate correction or the extra margin term. The K_{IR} curve is obviously very conservative. The fluences for the data shown in Fig. 4 are: 2.7×10^{19} n/cm^2 for the SA302B steel; 0.1, 0.7, and 2.5×10^{19} n/cm^2 for weld heat E19; and 0.7×10^{19} n/cm^2 for weld heat E23.

A key issue of concern is the effect of low upper shelf toughness on the shape of the dynamic and crack arrest fracture toughness data after irradiations indicative of 30 to 40 years of operation. Unfortunately, data reflecting these conditions investigated here do not resolve this concern. There is a definite need for more data, especially crack arrest, to validate the shifting K_{IR} curve approach. It is already known that the K_{IR} curve may not extend as high as 220 MPa $\cdot$ m$^{1/2}$ (200 ksi $\cdot$ in.$^{1/2}$) for low upper shelf material, since initiation fracture toughness (derived from J_{Ic} values) can be lower than this level. Similarly, the Charpy V-notch energy curve can change shape dramatically after irradiation (in addition to the shift in transition temperature and the drop in upper shelf), which suggests that the dynamic/arrest fracture toughness curve may also change shape to some degree. Studies on static fracture toughness curves after irradiation have

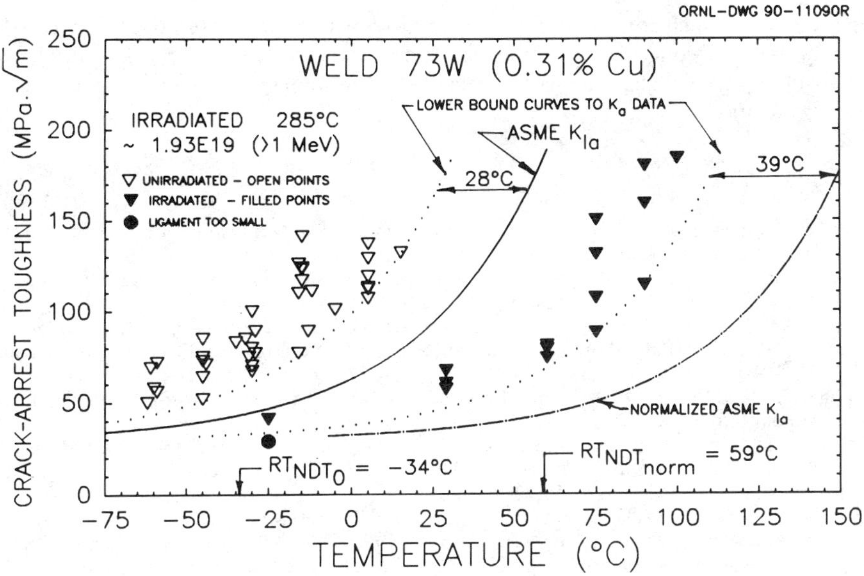

FIG. 5—*HSSI 6th Irradiation Series results for high-copper content weld 73W (data provided by Oak Ridge National Laboratory).*

shown some trends for a mean curve shape change. An approach using a different reference curve approach using a complex statistical lower-bound correlation also suggests a curve shape change for the static fracture toughness data [*9*].

Therefore, the concern for low upper shelf materials remains open. The Heavy Section Steel Irradiation (HSSI) program at Oak Ridge National Laboratory under the funding of NRC Research will provide significant information on low upper shelf welds in the 6th and planned 8th, 9th, and 10th Irradiation Series. The 6th Irradiation Series results were recently released [*12*], and the general conclusion is that the crack arrest toughness values produce conservative and consistent results with regard to measured fracture toughness and Charpy 41-J (30-ft-lbf) shifts. Figure 5 shows the 6th Irradiation Series results for the high copper content weld metal 73W; note the degree of conservatism between the K_{IR} curve and the actual crack arrest toughness data both before and after irradiation. The 10th Irradiation Series involves the testing of Linde 80 weld metal obtained from the cancelled Midland vessel. This particular weld metal is known to exhibit low upper shelf toughness, and the mechanical and fracture properties for this material before and after irradiation will be evaluated by several laboratories throughout the United States.

Conclusions

Neutron irradiation damage causes the fracture toughness of reactor pressure vessel steels to be diminished with increased exposure. The degree of embrittlement must be factored into operating limits for vessels by following the rules contained in the ASME Code and NRC Regulations. This study has looked at some of the sparce but limiting irradiated dynamic initiation and crack arrest fracture toughness data for a direct comparison with the K_{IR} curve and Regulatory shift methodology. The result is that the Regulatory approach is very conservative since no data fall very near the shifted K_{IR} curve. These comparisons have relied upon test reactor irradiations which are not necessarily indicative of actual vessel response. Future data are needed to address the concern for possible excess conservatism in the shifted K_{IR} curve method and of possible K_{IR} curve shape changes (especially for low upper shelf toughness materials) after significant irradiation.

References

[*1*] *PVRC Recommendations on Toughness Requirements for Ferritic Materials*, WRC Bulletin 175, Welding Research Council, New York, Aug. 1972.

[*2*] *Flaw Evaluation Procedures; Background and Application of ASME Section XI Appendix A*, EPRI NP-719-SR, Electric Power Research Institute, Palo Alto, CA, Aug. 1978.

[*3*] Shabbits, W. O., *Dynamic Fracture Toughness Properties of Heavy Section A533 Grade B Class 1 Steel Plate*, WCAP-7623, HSST Program Technical Report 13, Westinghouse Electric Corp., Pittsburgh, PA, Dec. 1970.

[*4*] Ripling, E. J. and Crosley, P. B., "Strain Rate and Crack Arrest Studies," paper presented at the HSST Fifth Annual Information Meeting, 1971.

[*5*] Davidson, J. A. et al., *The Irradiated Dynamic Fracture Toughness of ASTM A533, Grade B, Class 1 Steel Plate and Submerged Arc Weldment*, WCAP-8775, HSST Program Technical Report 41, Westinghouse Electric Corp., Pittsburgh, PA, Oct. 1976.

[*6*] Shabbits, W. O., Pryle, W. H., and Wessel, E. T., *Heavy Section Fracture Toughness Properties of A533, Grade-B, Class-1 Steel Plate and Submerged Arc Weldment*, WCAP-7414, HSST Program Technical Report 6, Westinghouse Electric Corp., Pittsburgh, PA, Dec. 1969.

[*7*] *Radiation Embrittlement of Reactor Vessel Materials*, Regulatory Guide 1.99, Revision 2, Nuclear Regulatory Commission, Washington, DC, May 1988.

[*8*] Guthrie, G. L., *Correlations Between Power and Test Reactor Data Bases*, NUREG/CR-5328, PNL-6793, Nuclear Regulatory Commission, Washington, DC, Feb. 1989.

[*9*] Server, W. L. and Caldwell, H. M., *An Approach for Predicting Reference Fracture Toughness in Irradiated Vessel Materials*, EPRI NP-5793, Electric Power Research Institute, Palo Alto, CA, May 1988.

[*10*] Mager, T. R. and Marschall, C. W., *Development of a Crack Arrest Toughness Data Bank for Irradiated Reactor Pressure Vessel Materials*, EPRI NP-3616, Electric Power Research Institute, Palo Alto, CA, July 1984.

[*11*] Hawthorne, J. R. et al., *Evaluation and Prediction of Neutron Embrittlement in Reactor Pressure Vessel Materials*, EPRI NP-2782, Electric Power Research Institute, Palo Alto, CA, 1978.

[*12*] Iskander, S. K., Corwin, W. R., and Nanstad, R. K., *Results of Crack-Arrest Tests on Two Irradiated High-Copper Welds*, NUREG/CR-5584, (ORNL/TM-11575), Oak Ridge National Laboratory, Oak Ridge, TN, Nov. 1990.

George R. Irwin,[1] *James W. Dally,*[1] *Xian-Jie Zhang,*[1]
and Robert J. Bonenberger[1]

Lower-Bound Initiation Toughness of A533-B Reactor-Grade Steel

REFERENCE: Irwin, G. R., Dally, J. W., Zhang, X-J., and Bonenberger, R. J., **"Lower-Bound Initiation Toughness of A533-B Reactor-Grade Steel,"** *Rapid Load Fracture Testing, ASTM STP 1130*, Ravinder Chona and William R. Corwin, Eds., American Society for Testing and Materials, Philadelphia, 1992, pp. 9–23.

ABSTRACT: The lower-bound initiation toughness of A533-B reactor-grade steel was determined over the temperature range from 3 to 50°C. The toughness of the steel was depressed toward the lower-bound value by using the following testing procedures: (1) dynamic loading, (2) notched-round-bar specimens, and (3) axial precompression of the notch. The paper describes in detail the method of applying the impact load to the specimen, the method of precompressing the specimens, and the testing procedure. The dynamic initiation toughness, K_{Id}, which correlates with the lower-bound toughness, was determined from analysis of the strain-time behavior of the specimens. Also, the results from a fractographic analysis were correlated with those from the strain-time analysis. The lower-bound toughness from this study compared favorably with K_{Ia} and K_{Id} data established through more extensive testing programs.

KEYWORDS: fracture, fracture toughness (lower-bound), impact testing, small specimen testing, steel (reactor-grade)

Nomenclature

a	Radius of the notched section, mm
a_e	Effective radius of the notched section, mm
b	Radius of the shoulder section, mm
d	Diameter of the notched section, mm
D	Diameter of the shoulder section, mm
K or K_I	Opening mode stress intensity factor, $MPa \cdot m^{1/2}$
K_{Ia}	Crack-arrest toughness, $MPa \cdot m^{1/2}$
K_{Ic}	Static initiation toughness, $MPa \cdot m^{1/2}$
K_{Id}	Dynamic initiation toughness, $MPa \cdot m^{1/2}$
r_Y	Exclusion adjustment for residual stress effects, mm
ϵ_o	Gross section strain, m/m
σ_o	Gross section stress, MPa
σ_n	Net section stress, MPa
σ_Y	Plastic flow stress, MPa
E	Young's modulus of elasticity, MPa

[1] Professor, professor, research associate, and graduate research assistant, respectively, Mechanical Engineering Department, University of Maryland, College Park, MD 20742.

$F_3(a_e/b)$	Numerical function
RT_{NDT}	Reference nil-ductility temperature, °C

Introduction

Crack-arrest toughness K_{Ia} indicates the K value below which the fine-scale cleavage events, even when initiated, are unable to spread and join and, thus, represents a lower-bound toughness. However, the current understanding of cleavage-fibrous behavior for nuclear reactor vessel steels suggests that a method of cleavage initiation testing with small specimens may provide the same lower-bound data with more efficiency. The behavior that handicaps slow-load, small-specimen testing to determine cleavage initiation toughness is the large amount of scatter observed in the test results [*1–3*]. The degree of scatter indicates that cleavage initiation along the crack front in a small specimen must be considered as a rare event. Only when the number of small specimen tests is large do the lowest observed values correspond with the toughness determinations from large specimens with long crack fronts. However, if the rare event feature is removed by notch embrittlement, rapid loading, and constraint, a good possibility exists for lower-bound determinations based on the failure of initiated cleavage elements to spread and join. Dally et al. [*4*] developed a testing procedure, using relatively small specimens, for determining the lower-bound initiation toughness of reactor-grade steels. The results show less scatter than slow-load testing methods, a valuable attribute relative to the number of specimens needed to obtain valid toughness values.

This paper presents rapid-load measurements of the cleavage initiation toughness of A533-B reactor-grade steel over a limited range of temperatures. The approach presented in Ref *4* is used in this study. A critical element in the success of small specimen determination of the lower-bound cleavage initiation toughness is to increase the severity of the local stress adjacent to the precrack. By increasing the stresses local to the crack front, it is possible to match more nearly the probability of cleavage initiation sites that occur in large specimens or in components with a long crack front.

The specimen used in this study is a circumferentially notched round bar that provides significant constraint with a large elevation of the flow stress. Impact loading also serves to elevate the flow stress caused by strain rate effects. Finally, a circular precrack, concentric with the round bar, is formed by axial compression. The axial compression closes a small segment at the root of the notch to form a pseudo crack. After release of the compressive load, a small natural crack is formed at the tip of this pseudo crack. The precompressive process also produces residual tensile stresses at the crack tip, which further elevates the stress to increase severity and enhance the probability of a lower-bound cleavage initiation.

The paper describes in detail the method of applying the impact load to the specimen. Also described is the method of precompressing the specimens and the testing procedure. Strain-time traces are included to characterize the response of the specimen to the impact loading. A method is presented for determining the dynamic initiation toughness, K_{Id}, corresponding to the lower-bound toughness, from these traces. An interpretation of the fracture behavior of the specimen based on the strain-time response is given. The results of a fractographic analysis are presented and correlated with the behavior observed from the strain-time traces. Finally, the results from this study are compared to the crack-arrest toughness K_{Ia} of the same material, established through independent extensive testing programs.

Notched-Round-Bar Specimen

The purpose of using a notched round bar as the specimen in a dynamic fracture initiation experiment is to simulate with a relatively small-diameter bar the constraint provided by a very

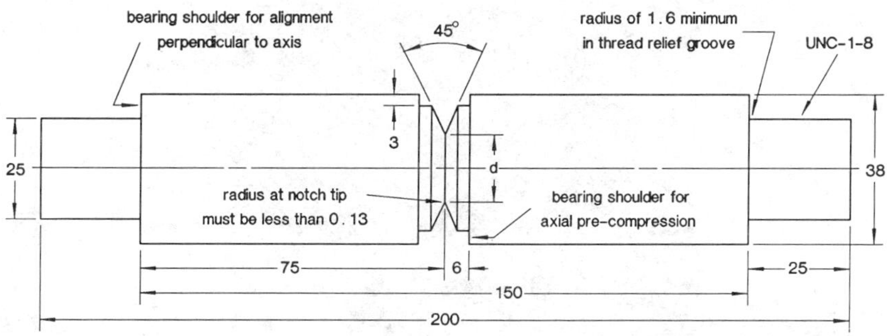

FIG. 1—*Dimensions of the notched-round-bar specimen.*

thick plate-type fracture specimen. A rigorous comparison of constraint afforded by a plate specimen of thickness, *B*, and a notched round bar with a shoulder diameter *D* has not been established; however, it is believed that the constraint provided by a notched round bar is at least equivalent to that provided by a plate specimen with a thickness equal to π times the net-section diameter, *d*.[2]

The round-bar specimen used in this initial study is illustrated in Fig. 1. The nominal shoulder diameter, *D*, is 38 mm, and the diameter, *d*, defining the notch section varied from 13 to 19 mm. The notch was machined with a 45° included angle, using a tool with a tip radius $r = 0.13$ mm. The specimen was shouldered at each end to provide accurately machined bearing surfaces perpendicular to the specimen axis. These bearing surfaces were essential to ensure alignment of the specimen in the impact loading device. Both ends of the bar were threaded (1–8 UNC) to provide a means for attaching the specimen to the loading train.

The total length of the bar was 200 mm, with 150 mm used for the center section. This section accommodated the notch while providing a uniform cross-sectional area of length $L = 2D$ above and below the notch. Shallow shoulders 3 mm deep are located above and below the notch. These shoulders are used to control the amount of permanent, axial deformation imposed on the notch during the initial precompression. Strain gages were mounted at a distance *D* above and below the notch to measure the nominal strains imposed on the specimen during impact loading.

Tensile Impact Loading Tower

The essential features of the loading tower are shown in Fig. 2 in which the delivery end of the system is illustrated. The impact load is developed by dropping a weight of 59 kg from a height that is adjustable from 0 to 1.8 m. The weight is fitted at its upper end with a brass bushing and at its lower end with a hardened steel insert (AISI 4340 with Rockwell C scale hardness $R_c = 48$). The bushing and insert both serve as bearings and keep the weight in alignment with the center rod.

The center rod, fabricated from drill stock 50 mm in diameter, is fitted with an internal 1–8 UNC thread at its lower end. This rod supports the top end of the round-bar specimen. A load transmission tube fits over the round-bar specimen overlapping the center rod by 25 mm. An anvil, threaded onto the other end of the specimen, supports the load transmission tube. All of the contacting surfaces on the loading fixture, namely the weight insert, the transmission tube,

[2] This estimate assumes the thickness of a plate specimen (*B*) is equivalent to the net-section circumference of the round bar, which results in the relation: $B = \pi d$.

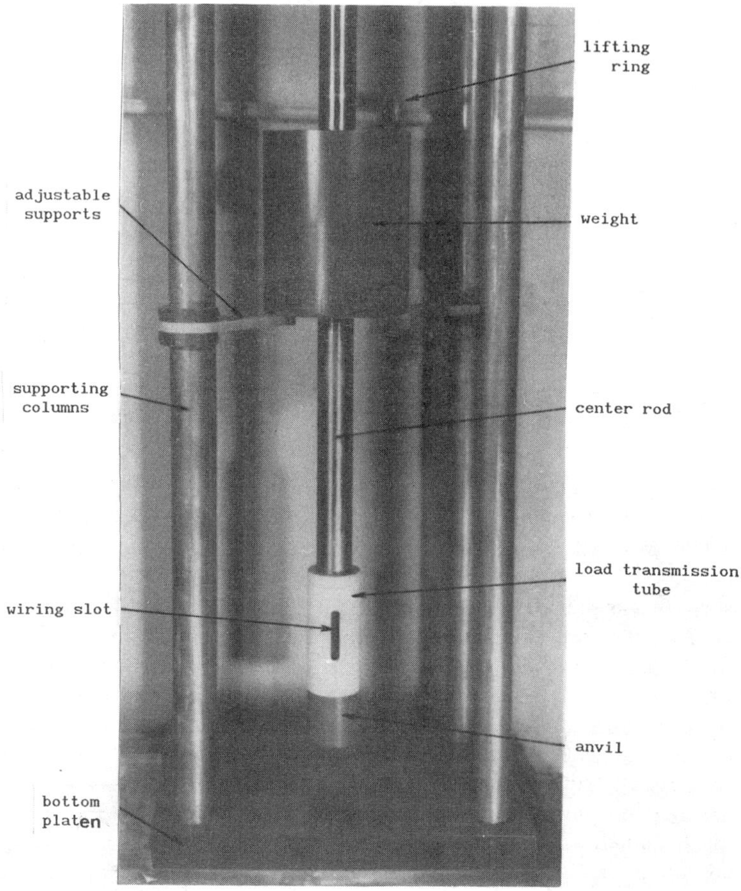

FIG. 2—*Key features of the tensile-impact-loading device.*

and the anvil, are flat, square to the axis of the center rod, and hardened to ensure axial impact and facilitate high loading rate.

The weight delivery system is supported by a four-column frame. Heavy steel plates 50 mm thick serve as the top and bottom platens. The center rod extends through the top platen and is positioned vertically with two hex nuts positioned on each side of the top platen.

The impact velocity of the weight with the maximum drop distance is 6 m/s. The energy delivered to the load transmission tube is 1053 N-m. However, the loading rate and the stress in the specimen depend mainly on the impact velocity rather than the available drop-weight energy. In addition, the stresses and the load rate depend on the geometry of the specimen and the yield characteristics of the material being tested. The load imposed on the specimen is measured directly during the impact period with strain gages.

Axial Precompression of Notch

Although the tip of the notch was machined with a radius of about 0.13 mm, it was not sufficiently sharp to represent a crack or initiate fracture during impact with the maximum

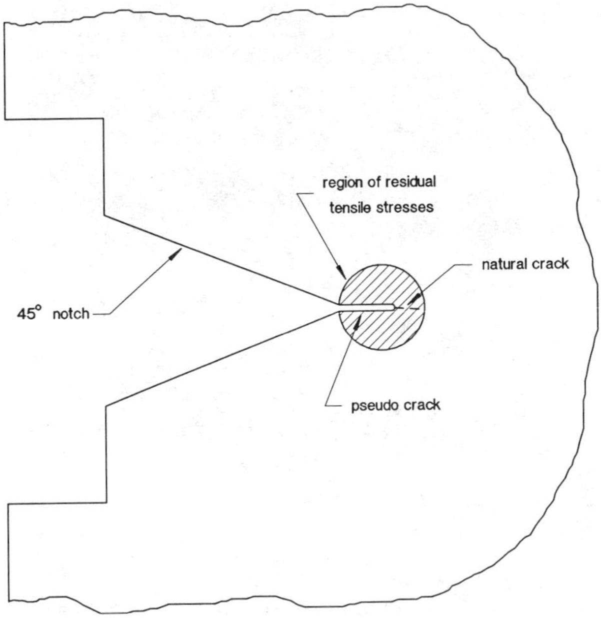

FIG. 3—*Pseudo crack formed from axial precompression.*

available drop-weight velocity. The crack was sharpened by applying an axial compressive load, which exceeded the uniaxial yield strength (483 MPa) of A533-B by a factor ranging from 3 to 7. Yielding at the center section of the bar caused the sides of the notch near the tip to move together to produce a pseudo crack (Fig. 3).

The straight crack shown in Fig. 3 is the pseudo crack formed by closing the two sides of the notch as the region local to the notch and, to a lesser degree, the central region both yield. A short extension occurs at a shallow angle to the pseudo crack. This extension is a natural crack, which is produced by the residual tensile stresses that develop as the axial load is removed from the bar.

The effect of the precompressive process on the mechanical properties of the steel is not known. Under the axial compressive load, the notched section of the bar is strained plastically to levels of 10 to 20%. With this large amount of plastic strain, one expects some degree of work hardening, except for the Bauschinger effect. Because the round bar is preloaded in compression and tested in tension, the Bauschinger effect should produce a reduction in the tensile yield strength and an apparent softening of the material in the notch section. It is believed these effects are small because the work hardening of these low-carbon steels at these strain levels is not significant.

In this method of crack sharpening, the compression of the notch should be uniform about the circumference. To facilitate uniform deformation, the round bar was fitted with end anvils with their bearing surfaces perpendicular to the axis of the bar. Also, a split V-groove fixture was clamped to the body of the bar to prevent any off-axis bending of the bar by the compressive load. The uniformity achieved is demonstrated in a ×5.5 fractograph (Fig. 4), in which the pseudo crack appears as a ring with a uniform thickness.

In addition to the circumferential uniformity of the deformation, the amount of axial compression must be carefully controlled. Enough axial deformation must occur to sharpen the

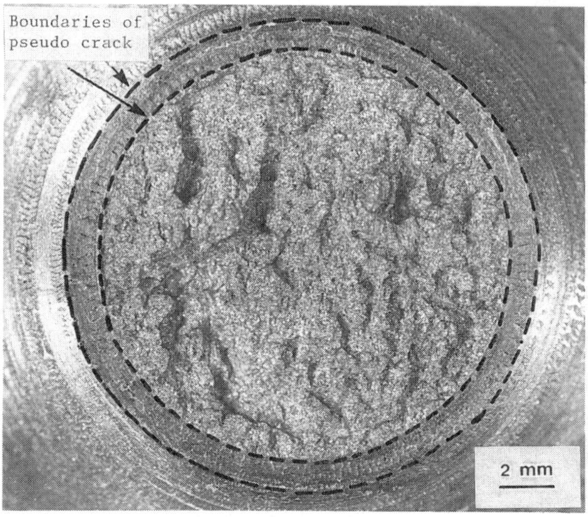

FIG. 4—*Optical photograph showing a ring of pseudo crack.*

notch tip and induce sufficient residual tensile stresses at the root of the notch to elevate the flow stresses. To control the axial deformation, steel spacing rings 3 mm thick were placed in the notch area shoulders (Fig. 5). The rings acted as mechanical stops, controlling the axial deformation so that it was uniform around the circumference. By using several rings with varying heights, different amounts of deformation were imposed. The compression load required was measured directly on the universal testing machine.

Test Procedure

Strain gages were used to determine the load imposed on the notched-round-bar specimens during impact. The gages were oriented in the axial direction and placed at 120° intervals

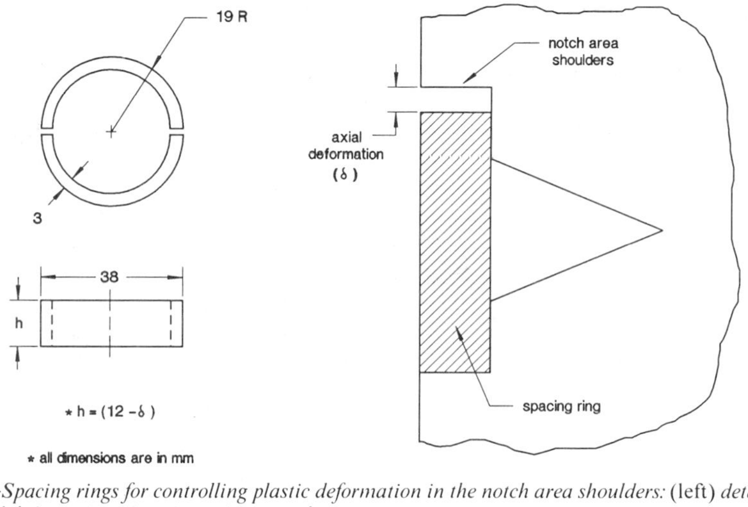

FIG. 5—*Spacing rings for controlling plastic deformation in the notch area shoulders:* (left) *detail of spacing rings and* (right) *spacing rings in position on the specimen.*

around the specimen at a distance of D above and below the notch. General purpose strain gages were employed with a nominal resistance of 350 Ω and a gage length of 1.6 mm. The gages were connected to a Wheatstone bridge/amplifier unit, capable of 100 kHz, for appropriate signal conditioning.

A type J, iron-constantan thermocouple, with a resolution of 0.1°C, was mounted in the notch area to measure the testing temperature of the specimen. Twenty specimens were tested over a range of temperatures from 3 to 50°C. Dry ice was used to cool the specimens, and a resistance heater, attached to the bottom anvil, was used to heat the specimens.

All of the tests were conducted using the maximum capacity of the loading frame, with a drop of 1.8 m for the weight. As the weight strikes the transmission tube, a compressive stress wave propagates through the tube and into the anvil. After encountering the bottom free surface of the anvil, the compressive pulse is reflected as a tensile pulse and propagates upward, into the notched-bar specimen. As expected, the strain gages mounted below the notch responded first, and the gages mounted above the notch responded about 15 μs later. This lag is due to the time required for the tensile stress wave, propagating at about 5 mm/μs, to travel the distance between the gages.[3]

The voltage output from one of the bottom strain gages was used to initiate the sweep of three digital storage oscilloscopes, allowing the voltage-time traces from each gage to be recorded with a common time base. The analog-to-digital converter on each of the oscilloscopes was set to sample at a rate of 200 ns/point. The voltage-time traces were then downloaded from the oscilloscope memories to a personal computer for further data processing.

Lower-Bound Fracture Initiation Toughness

The method of analysis to determine the dynamic initiation toughness K_{Id} from strain-time traces recorded from the round bar during impact testing is based on relations derived for static loading. For a notched, round bar subjected to uniaxial tension, the stress intensity factor is given by Tada et al. [5] as

$$K_I = \sigma_n \sqrt{\pi b}\, F_3(a_e/b) \tag{1}$$

where $F_3(a_e/b)$ is a function defined by

$$F_3(a_e/b) = (1/2)\sqrt{(a_e/b)[1 - (a_e/b)]}[1 + 0.5(a_e/b) + 0.375(a_e/b)^2 - 0.363(a_e/b)^3 + 0.731(a_e/b)^4] \tag{2}$$

Note also that

$$\sigma_n = (b/a_e)^2 \sigma_0 \tag{3}$$

The effective radius a_e of the net section is given by

$$a_e = a - r_Y \tag{4}$$

where r_Y is an exclusion adjustment to account for the effect of residual stress at the tip of the pseudo crack. The adjustment r_Y is expressed as

[3] For a more complete description of stress wave effects in the specimen, refer to Ref *4*.

$$r_Y = (1/2\pi)(K/\sigma_Y)^2 \tag{5}$$

In determining r_Y, the plastic flow stress, allowing for constraint and rapid loading, was assumed to be 1091 MPa, and K was the estimate of the stress intensity factor before the adjustment of the radius of the net section.

The flow stress was determined by first increasing the static yield strength of A533-B steel from 483 to 905 MPa, to account for the constraint. Next, strain-rate effects were considered by adding another 186 MPa, which is a common practice for structural steels. These adjustments for constraint and strain-rate effects gave a dynamic, plastic flow stress estimate of 1091 MPa.

For uniaxial loading, the strain measured on the shoulder section of the bar is

$$\epsilon_0 = \sigma_0/E \tag{6}$$

where E is 207 GPa for carbon steel. If the strain ϵ_0 in Eq 6 is the strain measured at initiation, then application of Eqs 1 through 5 give the dynamic initiation toughness, K_{Id}.[4]

The difference between the static and dynamic values of the stress intensity factor K and the J integral for notched round bars has been analyzed with finite elements by Nakamura et al. [6]. The dynamic loading was a tensile pulse that increased linearly with time for a 50-μs period. Nakamura et al. found that the J integral resulting from the tensile pulse was essentially the same as that produced by static loading of the same magnitude. The referenced study was performed to justify the data analysis method used by Costin et al. [7] in determining K_{Id} and J_{Id} for Hopkinson bar experiments. Because the round-bar experiments described here involve strain rates much lower than the strain rates produced in a Hopkinson bar, it appears that the static analysis described above is adequate for predicting the initiation toughness from the strain measurements made one diameter from the notch.

Test Results

Twenty specimens fabricated from A533-B steel were tested in axial impact over a range of temperatures from 3 to 50°C. Of this group, eleven specimens failed in a manner such that a valid (acceptable) K_{Id} value could be determined. The voltage-time data for each specimen were imported into a commercial spreadsheet program, and strain-time traces were generated for the strain gages. From a preliminary observation of the traces, the mode of failure could be determined. The round-bar specimens failed in one of two ways: (1) by cleavage with very small amounts of ductile tearing or (2) by extensive ductile tearing before cleavage initiation. The first mode yields data that permit a valid value of K_{Id} to be determined from Eqs 1 through 6, but the second mode does not. Examples of valid and invalid strain-time traces are shown in Figs. 6 and 7, respectively. For a valid trace, a single peak value of strain marks the failure of the specimen. The strain increases monotonically with time for about 100 to 140 μs after the stress wave reaches a bottom gage, and then the load decreases rapidly after failure initiates. For an invalid trace, the maximum strain is maintained for an extended period of time, partial unloading/reloading takes place in the specimen, and complete failure does not occur for several hundred microseconds, as indicated in Fig. 7. These observations of the strain-time traces allowed separation of specimens failing by brittle cleavage from those failing by excessive ductile tearing before cleavage initiation.

Several common features, illustrated in Fig. 8, occur in all of the strain-time traces yielding valid K_{Id} values. Three traces are shown: (1) the average of all the bottom gages, (2) the average

[4] Only one iteration was performed when determining the exclusion adjustment necessary for K_I.

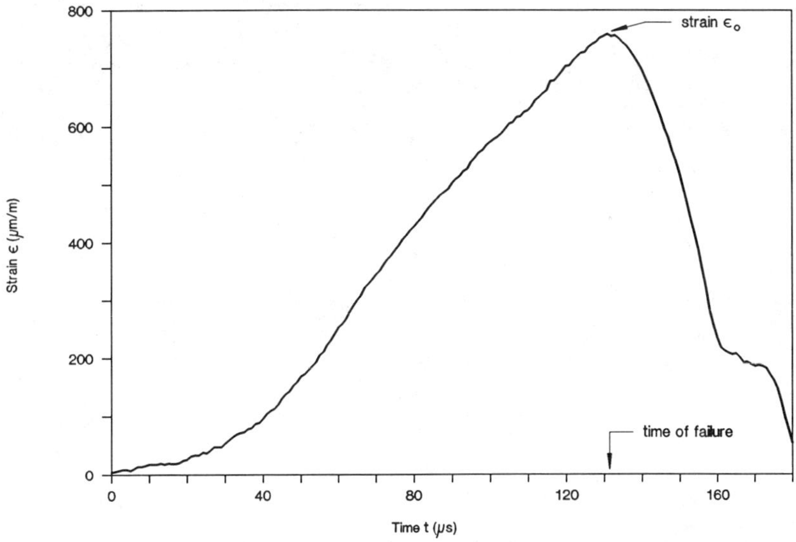

FIG. 6—*Strain-time trace for a valid test.*

of all the top gages, and (3) the average of all the gages. The bottom gages initially respond before the top gages because of the delay as a result of the time required for the stress wave to travel between the gage sets. Both sets of gages initially increase monotonically with the bottom gages registering higher strains than the top gages. This initial part of the fracture is dominated by stress wave behavior. At about 100 μs after the stress wave reaches the lower gages, the strain-time traces cross over, and the top gages indicate strains slightly higher than the bottom gages. It

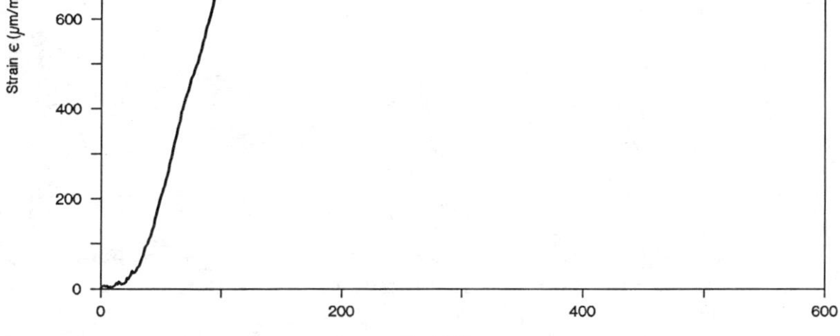

FIG. 7—*Strain-time trace for an invalid test.*

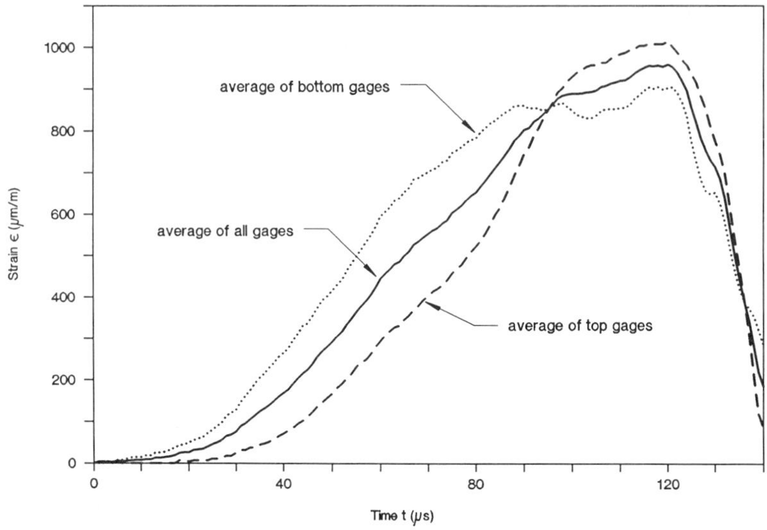

FIG. 8—*Common features of a valid strain-time trace.*

is believed that the crossover is due to stress wave reflection from the notch discontinuity, which suppresses the strain on the lower gages. After the crossover point, the traces from both gage sets correspond closely with the grand average until failure at 120 μs. This correspondence indicates that stress wave effects have diminished and that a quasi-static loading is prevalent. Both traces simultaneously record the strain at failure, which is the highest point on each trace. The simultaneity is expected because the notch is centrally located between the gages. For the time period between cross-over and failure, averaging methods provide a technique to determine the nominal strain ϵ_0 with a range of $\pm9\%$

The maximum strain from the strain-time traces was used as the failure strain ϵ_0 for the eleven qualifying tests. Values of K_{Id} were calculated for each of the six strains recorded. Two averaging methods were used to determine a total K_{Id} for each specimen. First, the peak values of the individual strain-time traces were averaged to compute K_{Id}. Then, all of the strain-time traces were averaged into a single trace, and K_{Id} was computed from the peak value of this combined trace. The difference in K_{Id} between these methods was only 4%, demonstrating the equivalency of both procedures. All subsequent K_{Id} values presented were determined by using the first method. The values of K_{Id} determined in this manner are shown as a function of temperature in Fig. 9.

Fractographic Analysis

Fractographic analysis is an essential component in verification that lower-bound values of initiation toughness have been achieved in the notched-round-bar test. In the determination of the lower-bound initiation toughness, one seeks to initiate cleavage with a minimum amount of prior ductile tearing (crack extension by hole formation and then hole joining). It is also important to initiate cleavage from several different sites distributed randomly about the circumference of the sharpened notch. Finally, the initiation sites should all be activated at nearly the same time or at the same load. Analysis of the fracture surface in a scanning electron microscope

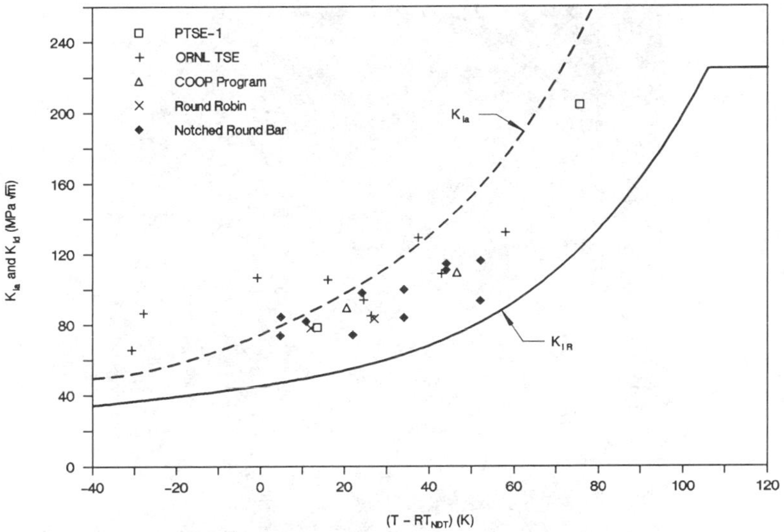

FIG. 9—*Comparison of* K_{Id} *and* K_{Ia} *test results for A533-B reactor-grade steel.*

(SEM), with a magnification of about ×9, shows several surface characteristics that verify the adequacy of the lower-bound determination.

An example of the surface features for a fractured round-bar specimen that did not provide a valid lower-bound determination is presented in Fig. 10. The crack extended by hole joining in a ductile tearing mode. The crack extension by ductile tearing varied around the circumference

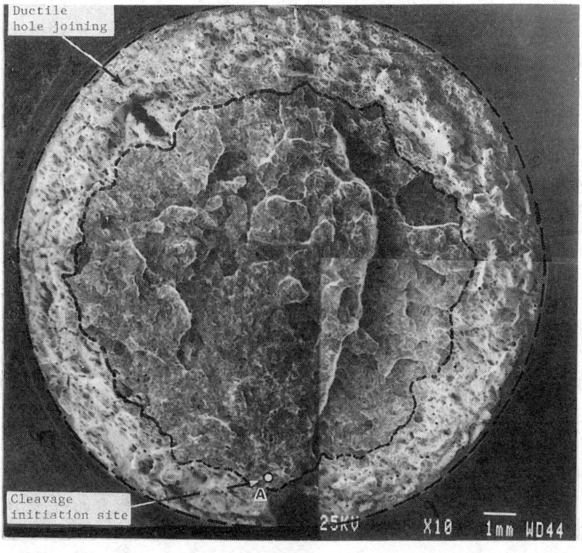

FIG. 10—*Fracture surface exhibits an extensive ring where crack extension occurred by hole joining before cleavage initiation.*

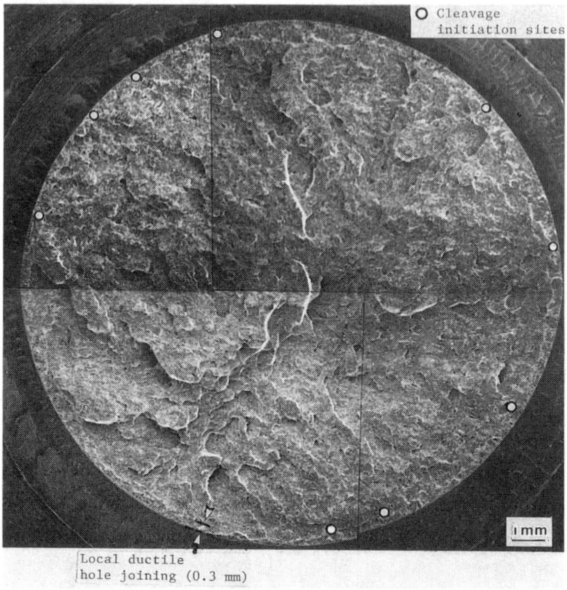

FIG. 11—*Fracture surface exhibits very limited hole joining. Initiation is at multiple sites about the circumference, and extension is dominated by cleavage.*

from a minimum of 1.8 mm to a maximum of 3 mm. The crack extension by tearing occurred at a low velocity while the specimen remained under dynamic load for more than 600 μs (see Fig. 7). The crack extension underwent a transition from ductile to brittle at an initiation site located at Point A. The crack then extended at high velocity ($\approx$500 m/s) over the central region of the specimen in a mode that was predominantly cleavage. Even in the central region, ridges are observed, indicating some areas of fibrous failure between cleavage regions.

Figure 11 illustrates a valid lower-bound determination. At the outer edge of the fractured specimen, a ring with a uniform thickness is evident, which indicates that the axial precompression was performed with controlled alignment. Next, the extension of the crack by ductile hole joining was limited from 0.3 mm at the crack front in the third quadrant to only 5 to 10 μm around the crack front in the first and fourth quadrants. Crack initiation occurred at about ten sites distributed nearly randomly about the circumference. Most initiation sites were located in regions along the crack front where the ductile tearing was minimized. The crack propagated at high speed ($\approx$500 m/s) across almost the entire test section. The specimen failed in 100 μs with a strain-time record similar to that shown in Fig. 6. This strain-time trace shows a well-defined peak, with no evidence of the plateau that is evident when extensive ductile tearing occurs.

The final example (Fig. 12) shows a borderline determination of the lower-bound toughness. The test was considered valid, and the data point was plotted in Fig. 9; however, the validity of the test can be debated. The fractograph clearly shows initial crack extension by hole joining that varied from 0.2 to 0.9 mm. Cleavage was initiated at a limited number of sites, and high-speed crack propagation took place from the bottom portion of the fracture area, at Site A, toward the top of the area. The strain-time traces also showed the effect of the low-velocity crack extension by ductile hole joining. The time required to fail was about 170 μs, which was long compared to many other valid tests. Also, the strain-time trace (Fig. 13) exhibited a 40-μs plateau, which is

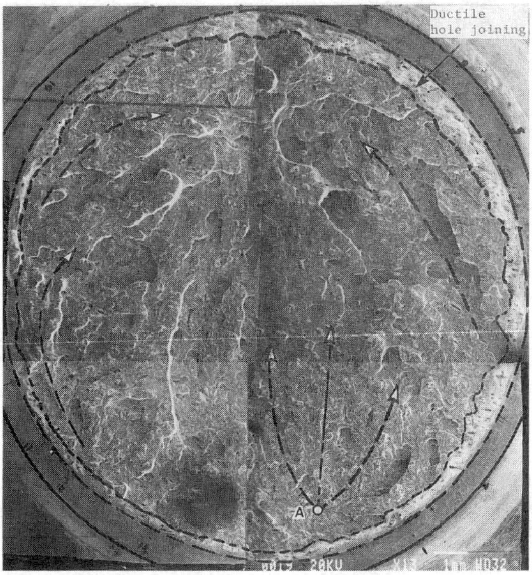

FIG. 12—*Fracture surface exhibits a moderate amount of crack extension by hole joining before cleavage initiation from a limited number of initiation sites.*

indicative of the load relief produced by low-velocity crack extension over a significant region of the fracture surface.

The amount of crack extension by ductile hole joining that can be tolerated in a lower-bound determination is still an open issue. The tearing relieves constraint, blunts the crack, and elevates

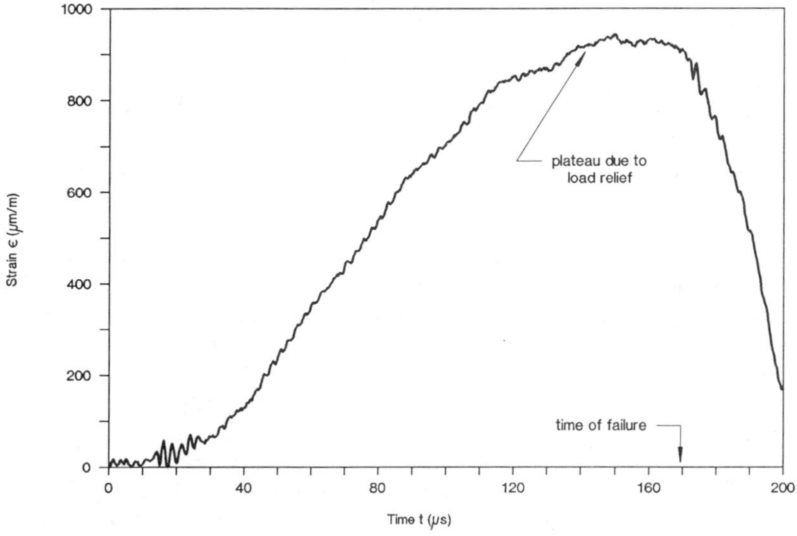

FIG. 13—*Strain-time trace for a borderline test.*

the apparent toughness. For these reasons, the amount of tearing permissible in a lower-bound determination must be limited. These early results indicate that ductile crack extension of 0.9 mm may be excessive.

Discussion

Results for the initiation toughness for A533-B steel over the temperature range from 3 to 50°C are presented in Fig. 9. Also included in this figure are data from K_{Ia} testing of the same material.[5] Data from eleven valid specimens show a somewhat smaller scatter band than the root mean square (rms) scatter obtained in crack-arrest testing.

The significant value of K from the test is the K value pertaining to the spreading of cleavage from regions where cleavage has initiated. Obviously, the net section reduction by the pseudo crack formed during precompression, as well as any small amounts of hole-joining fracture, are excluded from the net-section diameter. Another allowance for the cleavage initiation region was the reduction of the net-section diameter by $2r_Y$ to accommodate for the effects of residual stress at the crack tip.

The lower-bound initiation toughness, determined with the round-bar test procedure, compared closely with the crack-arrest toughness determined in the round robin and COOP program evaluation of A533-B reactor-grade steel [*8*]. As seen in Fig. 9, the tendency of previous K_{Ia} determinations was to indicate a lower-bound toughness moderately less than the crack-arrest toughness measured in tests of large specimens. This tendency seems to be matched by the results of our rapid load notched-round-bar experiments.

The amount of precompression required to minimize ductile tearing is not known. The amount of axial precompression imposed on a specimen is determined from experience and observation of the degree of tearing observed in invalid tests. Systematic study of the effect of the amount of axial precompression on both lower-bound toughness and the extent of crack extension by hole joining and by cleavage initiation is needed. The major test improvement needed, however, is an increase in the impact velocity so that larger net-section diameters can be used.

Conclusion

The notched-round-bar test procedure provides a relatively inexpensive method to determine lower-bound initiation toughness with small scatter. Measurements of the lower-bound toughness were made with A533-B reactor-grade steel, a relatively tough material, to a temperature of 50°C. The lower-bound toughness varied from 74 to 93 $\text{MPa} \cdot \text{m}^{1/2}$ as the test temperature was increased from 3 to 50°C.

Acknowledgments

The authors would like to thank Claude E. Pugh and William R. Corwin, of Oak Ridge National Laboratory, for providing support and encouragement in monitoring this research program. Thanks are also due to Dr. Donald B. Barker for his assistance in conducting some of the experiments.

References

[*1*] Iwadate, T., Tanaka, Y., Ono, S., and Watanabe, J., "An Analysis of Elastic Plastic Fracture Toughness Behavior for J_{Ic} Measurements in the Transition Region," in *Elastic Plastic Fracture: Second Symposium,*

[5] Note that the temperature in Fig. 9 is relative to the RT_{NDT}. For A533-B steel, $RT_{NDT} = -2°C$.

Vol. II: Fracture Curves and Engineering Applications, ASTM STP 803, C. F. Shih and J. P. Gudas, Eds., American Society for Testing and Materials, Philadelphia, 1983, pp. 531–561.
[2] Merkle, J. G., "An Examination of the Size Effects and Scatter Observed in Small-Specimen Cleavage Fracture Toughness Testing," ORNL Report TM-9088, NUREG CR-3672, 1984.
[3] Landes, J. D. and Shaffer, D. H., "Statistical Characterization of Fracture in the Transition Region," in *Fracture Mechanics, ASTM STP 700,* American Society for Testing and Materials, Philadelphia, 1980, pp. 368–382.
[4] Dally, J. W., Zhang, X. J., and Irwin, G. R., "Determining Lower Bound Dynamic Initiation Toughness from Notched Round Bars," in *Dynamic Fracture Mechanics for the 1990s, PVP—Vol. 160,* H. Homma, D. A. Shockey, and G. Yagawa, Eds., American Society of Mechanical Engineers, New York, 1989, pp. 47–54.
[5] Tada, H., Paris, P., and Irwin, G., *The Stress Analysis of Cracks Handbook,* Del Research, Hellertown, PA, 1973, p. 27.1.
[6] Nakamura, N., Shih, C. F., and Freund, L. B., "Elastic-Plastic Analysis of a Dynamically Loaded Circumferential Notched Round Bar," Materials Research Laboratory Report, MRL E-152, Brown University, Providence, RI, March 1984.
[7] Costin, L. S., Duffy, J., and Freund, L. B., "Fracture Initiation in Metals Under Stress Wave Loading Conditions," in *Fast Fracture and Crack Arrest, ASTM STP 627,* G. T. Hahn and M. F. Kanninen, Eds., American Society for Testing and Materials, Philadelphia, 1977, pp. 301–318.
[8] Barker, D. B., Chona, R., Fourney, W. L., and Irwin, G. R., "A Report on the Round Robin Program Conducted to Evaluate the Proposed ASTM Standard Test Method for Determining the Plane-Strain Crack-Arrest Toughness, K_{Ia}, of Ferritic Materials," ORNL Report NUREG/CR-4996, ORNL/SUB/79-7778/4, 1988.

DISCUSSION

Mark T. Kirk[1] (*written discussion*)—The notched-round-bar impact test proposed by the authors appears to have great potential as a dynamic fracture initiation test. Any comments the authors could provide regarding the possibility of performing this test in the old-style Charpy-tensile testing jig would be most helpful to other researchers attempting to perform this test using standard laboratory equipment.

Robert J. Bonenberger (*author's closure*)—During the initial stages of the test program, the feasibility of using the Charpy-tensile testing jig to load the notched-round-bar specimens was considered. However, the size of the specimens (D = 38 mm) prevented the use of the fixture. Although it may be possible to test subsize round-bar specimens in the Charpy machine, we do not recommend this practice because of the loss of constraint for the smaller specimen. Recently, the authors developed an alternate method for determining lower-bound initiation toughness of reactor-grade steels by using a modified form of a standard Charpy V-notch specimen, which can be tested in a standard Charpy impact machine. A paper describing the new method is being prepared and will appear in a future ASTM publication.

[1] David Taylor Research Center, U.S. Navy, DTRC Code 2814, Annapolis, MD 21146.

Hervé Couque,[1] *Robert J. Dexter,*[2] *and Stephen J. Hudak, Jr.*[1]

Using Small Specimens to Measure Dynamic Fracture Properties of High-Toughness Steels

REFERENCE: Couque, H., Dexter, R. J., and Hudak, S. J., Jr., **"Using Small Specimens to Measure Dynamic Fracture Properties of High-Toughness Steels,"** *Rapid Load Fracture Testing, ASTM STP 1130,* Ravinder Chona and William R. Corwin, Eds., American Society for Testing and Materials, Philadelphia, 1992, pp. 24–36.

ABSTRACT: The use of coupled pressure bars (CPB) to induce dynamic fracture in tough materials using small specimens is investigated. CPB experiments were performed with a nuclear pressure vessel steel, A533 Grade B Class 1, over the temperature range 37 to 100°C. The dynamic fracture initiation toughness at a stress intensity loading rate, $\dot{K}_I$, of 2×10^6 MPa · m$^{1/2}$ s^{-1} was deduced from the simulation of the fracture experiment with a dynamic viscoplastic finite-element fracture code. At 100°C, no cleavage fracture was observed for either dynamic crack initiation or subsequent propagation and arrest. A procedure to measure initiation, propagation, and arrest toughnesses of nuclear pressure vessel steels up to service temperature with CPB specimens is introduced.

KEYWORDS: dynamic fracture toughness, crack arrest, upper-shelf fracture toughness, A533-B steel, pressure vessel steel

To prevent catastrophic failure of engineering structures, fracture properties have been measured under dynamic loading conditions to consider the effect of material inertia as well as rate sensitivity. One major application is the integrity of nuclear pressure vessels. The characterization of the dynamic fracture properties of pressure vessel steels from room temperature to service temperature (320°C) presents an experimental challenge. This is due to the increase of the dynamic toughness in the transition regime to toughnesses of 200 to 400 MPa · m$^{1/2}$ in the temperature range 23 to 80°C [*1*]. These values exceed the already high upper-shelf static toughness. Until recently, experimental technology involving laboratory specimens has successfully measured toughnesses up to 230 MPa · m$^{1/2}$ [*2–4*]. Beyond 230 MPa · m$^{1/2}$, large specimen fracture tests have been the only reliable experimental approach [*1*]. Another challenge related to nuclear pressure vessel steels is the identification of a reliable procedure for the dynamic testing of small coupons from surveillance capsule programs. A need exists to develop techniques to characterize high-toughness materials using small specimens, not only for economic reasons, but also for situations in which only a limited quantity of material is available. The present investigation reports the current limitations in small specimen testing of tough materials and consequent remedy achieved with a new experimental technique involving coupled pressure bars (CPB).

The initiation of unstable crack propagation requires the rapid release of a critical amount of elastic strain energy to the crack tip. For tough materials, conventional technology achieves this

[1] Senior research engineer and manager, Mechanics of Materials, respectively, Engineering and Materials Sciences Division, Southwest Research Institute, 6220 Culebra Rd., San Antonio, TX, 78228.

[2] Senior research engineer, Advanced Technology for Large Structural Systems, Lehigh University, 117 ATLSS Dr., Bethlehem, PA 18015.

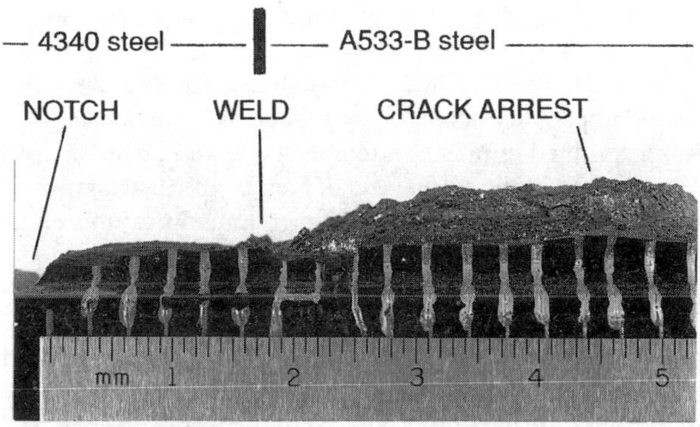

FIG. 1—*Crack profile of a side-grooved 4340 steel/A533B steel duplex specimen tested at 23°C.*

condition through the storage of strain energy in the specimen [5]. A machined notch within a compact specimen is loaded with a wedge and split pin assembly. The higher load required to initiate a machined notch versus a prefatigued crack enables the storage of a large amount of strain energy in the specimen. Such an experimental approach precludes the dynamic loading of a prefatigued crack. Using this technique, rapid crack propagation has been achieved with high-toughness pressure vessel steels of nil-ductility transition temperature, −20 to −27°C, tested up to a temperature of 23°C [4,6]. With increased temperature, a greater amount of strain energy is required to promote rapid crack propagation. This is associated principally with the increase of toughness due to the failure mode evolving from a cleavage (low-toughness) to a fibrous (high-toughness) type of failure. With larger notch radii, larger amounts of energy or preload can be obtained. However, because of the low flow stress of pressure vessel steels and the associated high toughness of these materials, large deformation and stable tearing occur, thereby precluding rapid, unstable fracture. Consequently, larger specimens similar in size to specimens of full-scale experiments would be required to investigate the transition regime and upper-shelf regime with these conventional approaches.

To increase the amount of stored energy within a given specimen size without causing large deformation or stable tearing, a modified wedge-loaded specimen, termed a *duplex specimen*, was introduced [7]. This specimen consists of a high-strength steel, containing a machined notch, which is welded to the high-toughness material to be tested. A greater load is required to initiate the dynamic event at the notch in the high-strength steel as a result of the high yield stress and moderate toughness. The rapidly propagating crack penetrates the high-toughness material. The larger amount of strain energy stored in the specimen can promote the rapid fracture propagation in A533 Grade B Class 1 (A533B) steel up to temperatures of 50°C [3]. Duplex specimens are attractive because of their intermediate size (W = 127 mm) and their low operative cost. However, as toughness increases, complications occur during the rapid crack propagation event at the weld between the high-strength steel and the high-toughness material. Specifically, the specimen often exhibits a nonplanar crack front in the high-toughness material, which originates from the weld zone. The crack front is traveling in a layer 15 mm high; this deviation is comparable to the plastic zone size. Such crack profiles have been observed in a series of experiments performed at 23 and 37°C with duplex, side-grooved specimens of A533B steel. This steel originated from the same heat as a series of wide-plate crack-arrest specimens, WP-1 [1]. This is illustrated in Fig. 1 for one of the specimens tested at 23°C. The analysis of such an experiment with the conventional experimental fracture mechanics approach is highly uncertain due to the diffuse nature of the crack front, as well as the out-of-plane growth pattern. However, the material properties characterizing such diffuse cracks need to be established. Such phenomena have been observed in full-scale experiment over a larger scale when compared to the duplex specimens (see Fig. 1 and Ref *1*). Having a small specimen technology precluding propagation of diffused cracks within the transition and upper-shelf temperature regime may provide a lower-bound toughness.

For material of lower nil-ductile transition temperatures and upper-shelf Charpy energy than the A533B steel of Ref *1*, such as the A508 steel of Ref *2*, rapid crack growth has been observed within the side-grooved plane of the compact specimen to temperatures of 23°C. However, this duplex technology seems to be limited to a toughness of 175 MPA $\cdot$ $m^{1/2}$ [*2*].

To circumvent the above limitations, a new technique has been introduced using two pressure bars to store elastic energy external to two precracked compact specimens [*8*]. The controlled fracture of an embrittled material starter specimen coupling the two pressure bars is employed to achieve a rapid release of energy from the pressure bars to the rigidly attached specimens.

In the following sections, the potential of the coupled pressure bars (CPB) technique for characterizing dynamic fracture properties of a nuclear pressure vessel steel, A533B, is presented. This steel originated from the same heat as the series of wide-plate crack-arrest specimens, WP-1 [*1*]. First, the experimental procedure applied to the testing of high-toughness materials is described. The CPB technique is then evaluated by performing tests in the 37 to 100°C temperature range. Finally, the benefits of this technology in characterizing the dynamic upper-shelf behavior of high-toughness steels are discussed.

Coupled Pressure Bars Experiment

The application of the CPB experimental procedure to high-toughness materials is introduced here. For the background on the design and development of the experimental apparatus, the reader is referred to Ref *8*.

A schematic of the CPB experiment is shown in Fig. 2. The primary components are two pressure bars to store energy, a starter specimen to release rapidly the stored energy, and two prefatigued, compact fracture specimens. The present experiments were conducted by preloading the pressure bars and starter specimen to 444 kN for testing temperatures less than or equal to 50°C and to 622 kN for higher testing temperatures. The test specimens were then inserted into slots in the bars and secured with wedges, as shown in Fig. 2. Fracture of the starter specimen was subsequently initiated by introducing a sharp cut into the circumferential notch of the starter specimen using a cutter wheel and high-speed air drill. Failure of the starter specimen releases an unloading stress wave in the bars, which transmits a rapid axial displacement rate to the specimens. The specimen crack length was chosen such that crack initiation occurred during the failure of the starter. During this time period, a monotonic load is applied to the specimen arms corresponding to a constant stress intensity rate, $\dot{K}_I = 2 \times 10^6$ MPa $\cdot$ $m^{1/2}$ s^{-1}. The subsequent rapid crack propagation event then occurs under a constant crack-opening displacement rate at the specimen load-line location, $C\dot{O}D_{LL}$, caused by the final unloading of the bars:

$$C\dot{O}D_{LL} = 2C_o\sigma/E \tag{1}$$

where

C_o = the sound velocity,
E = Young's modulus, and
σ = the applied stress in the pressure bars.

Applied loads of 444 and 622 kN in the maraging steel bars correspond to a $C\dot{O}D_{LL}$ of 20 and 28 m s^{-1}, respectively.

Three strain gages were mounted on the upper bar 127 mm from the starter section. From the

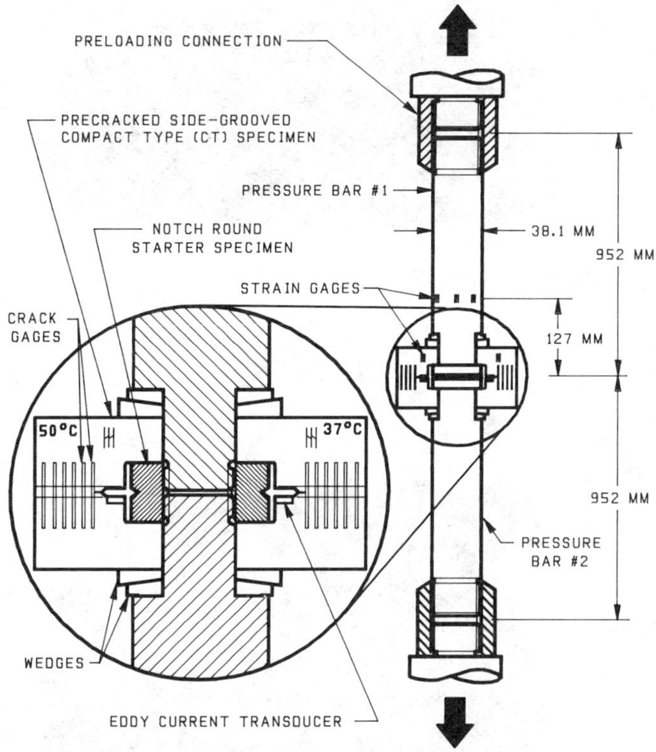

FIG. 2—*Schematic diagram of the coupled pressure bars experiment.*

strain-gage records, the failure duration of the starter (rising part of the pulse), as well as the duration of the constant displacement-rate regime (zero strain amplitude in the bar) was identified. In addition, one strain gage was mounted on each specimen 10 mm over the fatigue precrack tip location. The crack initiation time was deduced from this strain gage record by identifying the unloading compressive wave resulting from the initiation of the prefatigued crack. The crack-opening displacement history, COD_X, of each specimen was monitored at a distance X = 12.5 mm from the load line using eddy-current transducers attached to the specimen arms, as shown in Fig. 2. Crack propagation history $a(t)$ was monitored in each specimen using a ladder-type gage having six lines, spaced 3 mm apart. The ladder-type crack gage technique has been proven to provide a precise measure of the surface crack position based on calibrations performed with the optical method of caustics and with electrical strain gage measurements [*9*]. In this work, strain-gage and crack-gage responses were compared with A533B duplex specimens tested at 23°C. From the strain record, the arrival time of the crack at a given crack-gage location was deduced. This time was found to coincide within 2 μs to the failure time of the crack gage. It is believed that such technology gives the average position of the propagating crack front as a result of the limited shear lips formation occurring in these small, prefatigued side-groove specimens.

Results and Analysis

The feasibility of the new technique was evaluated with A533B steel from the same heat of material as that used for the series of wide-plate crack-arrest experiments, WP-1 [*1*]. Two prefa-

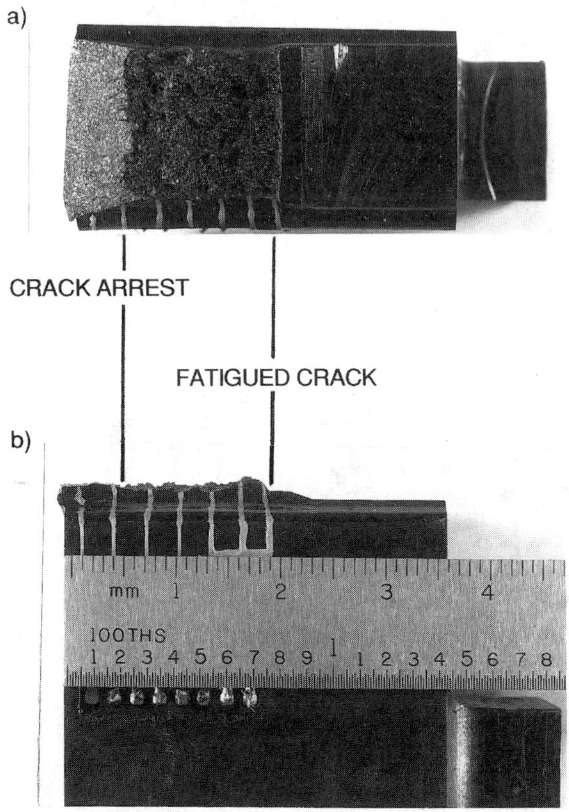

FIG. 3—*A533B specimen tested at 100°C:* (a) *fracture surface and* (b) *crack profile.*

tigued compact specimens were used for each test of planar size $W = 44$ mm and crack length $a_0 = 24$ mm. Two thicknesses, B, were used, specifically, 15 mm for the tests performed at 37 and 50°C and 20 mm for the tests performed at 75 and 100°C. The specimens were side-grooved to 25% of the thickness, B, resulting in net thicknesses, B_N, of 11.4 and 15.0 mm, respectively. Planar crack growth was obtained for each experiment, with deviations less than 1 mm over a distance of 14 mm or greater. This is illustrated in Fig. 3 with the specimen tested at 100°C.

Dynamic viscoplastic finite-element simulations of the CPB specimens were performed with a special-purpose, two-dimensional computer program, VISCRCK [*10*]. A Bodner-Partom constitutive model developed from tensile data obtained at strain rates varying from 10^{-3} to 3×10^3 s^{-1} over the temperature range −60 to 175°C was incorporated to the dynamic fracture code [*11*]. A mesh composed of 387 elements having linear dimensions of 2 mm was used.

The measured crack-growth history and estimated load-line deflection history from the strain-gage measurements on the pressure bars are input to the dynamic simulation. The specimen loading history was estimated based on strain measurements performed on the arms of a 4340 steel specimen tested at a preload of 444 kN. The input load history was taken to be increasing monotonically from 0 to 5% of the preload during the failure of the starter. Subsequently, during the final unloading of the pressure bars, the load was held constant and equal to 5% of the preload. These numerical simulations enable a comparison of calculated versus experimental values of the crack-opening displacement, COD_X, and strains.

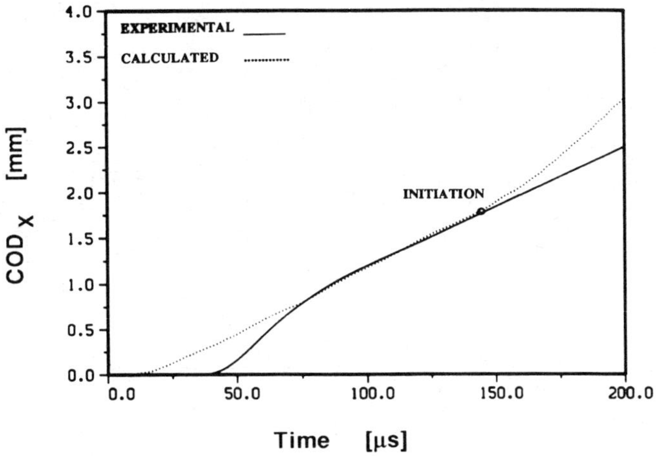

FIG. 4—*Comparison of measured and calculated crack-opening displacement records* $COD_X(t)$ *for the A533B specimen tested at 100°C.*

Two fracture criteria for ductile materials, the dynamic J' integral [*12,13*] and the T^* integral [*14*], were considered. The J' and T^* integrals are equivalent for monotonic loading up to crack initiation. The T^* integral is an incremental formulation proposed to handle the effects of viscoplasticity and unloading that occur during crack propagation. From the evaluation of the crack-tip integral, the stress intensity factor, K_I, is deduced by considering the small scale yielding relation under plane strain conditions

$$K_I = \sqrt{EJ'/(1 - \nu^2)} \quad (2)$$

where E and ν are Young's modulus and Poisson's ratio, respectively. The same relation is also used with T^*.

Figure 4 compares the computed crack-opening displacement history, COD_X, with the experimental data for the specimen tested at 100°C. Good agreement was obtained between the computed and experimental values up to initiation. Both crack-tip integrals, J' and T^*, were calculated and found to coincide during the process of crack initiation. Consequently, the stress intensity factor was deduced using Eq 2. As shown in Fig. 5, a dynamic fracture initiation toughness of 398 MPa $\cdot$ m$^{1/2}$ was obtained. This procedure was repeated for each specimen over the temperature range 37 to 100°C, and the results are summarized in Table 1.

The validity of the results was evaluated based on an adaptation of the static Paris' criterion to dynamic loading conditions [*15*]:

$$a_0, b, B \geq \alpha J_{Id}/\sigma_{yd} \quad (3)$$

where a_0, b, B, and σ_{yd} are the initial crack length, the remaining ligament, the thickness, and the dynamic yield stress, respectively, and $\alpha = 25$ (see Refs *16* and *17* and ASTM Test for J_{Ic}, a Measure of Fracture Toughness [E 813]). The relevance of using this criterion for dynamic ductile fracture has been recently demonstrated by Moran et al. [*18*]. The strain rate corresponding to the dynamic yield stress of Relation 3 was taken to be the average strain rate reached at the plastic zone boundary, defined at 0.002 strain. This strain rate was estimated using Costin et al.'s [*19*] approach and calculated to be 30 s^{-1} for A533B steel.

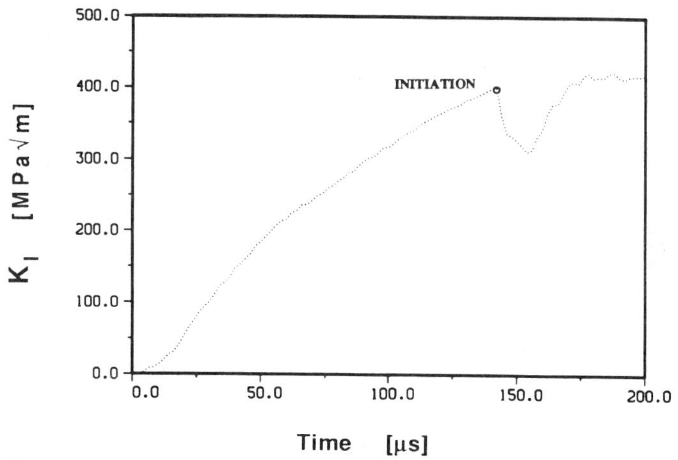

FIG. 5—*History of the crack driving force,* K_I, *for the A533B specimen tested at 100°C.*

For data not satisfying Relation 3, the overestimation of the toughness was deduced using the procedure developed by Couque et al. [*20*] based on the data of Landes and Begley [*16*] for a steel of similar flow stress and static toughness to the A533B steel. Using dynamic yield stresses at a strain rate of 30 s^{-1} over the temperature range of interest (see Ref *11*), the values of α, and eventual toughness corrections, were calculated and are reported in Table 1. Also indicated in Table 1 is a margin of error for the corrected toughnesses deduced from Landes and Begley's data. For A533B steels tested at 100°C, Relation 3 will be verified with CPB specimens 40 mm thick (B). An estimate of the specimen thickness to satisfy Relation 3 for a testing temperature of 320°C is discussed in the next section.

Discussion

The demonstration of the CPB technique as a useful technique for characterizing dynamic fracture properties of high-toughness steels is discussed in this section. The dynamic fracture properties and corresponding fracture morphologies are also compared with static fracture and arrest properties of A533B steel.

TABLE 1—*Dynamic initiation fracture toughness results.*

Testing Temperature, °C	Fibrous Mode,[a] %	K_{Id} Uncorrected, MPa · m$^{1/2}$	σ_{yd}, MPa	α	B_N, mm	K_{Id} Corrected, MPa · m$^{1/2}$	$\dot{K}_I$, 10^6 MPa · m$^{1/2}$ s^{-1}
37	30	220	501	27	11.5	220	2.2
37	30	180	501	40	11.5	180	2.0
50	85	250	491	21	11.5	228 ± 8	2.4
75	100	262	480	24	15.0	262	2.1
100	100	398	476	10	15.0	337 ± 16	2.8

[a] Percentage of initiation zone, defined as region within one calculated crack-tip opening displacement using procedure outlined in Ref *20* and A533B dynamic tension data [*11*].

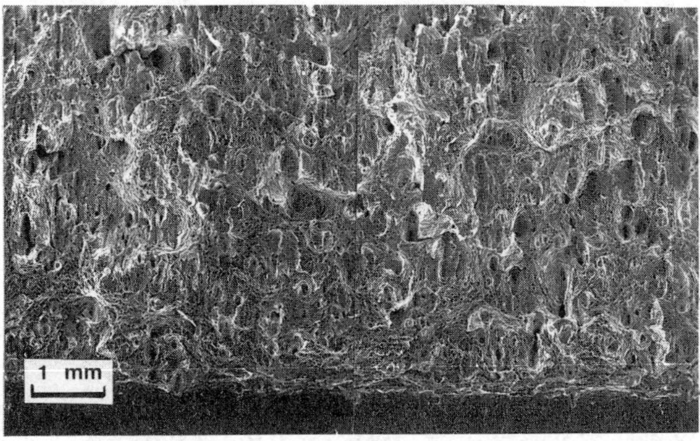

FIG. 6—*Scanning electron fractograph adjacent to the prefatigued crack tip of the A533-B specimen tested at 100°C. The crack velocities reached during the early stage of the rapid crack propagation event are indicated.*

The dynamic upper-shelf regime of the A533B steel was observed to be reached at 100°C. As shown in Fig. 6, no cleavage was noticed at this temperature, either at initiation or during propagation. Crack velocity up to 1000 m s^{-1} was reached with a fibrous type of failure (Fig. 7). The higher crack velocity reached during the early crack propagation event is related to the initiation event. The release of a large amount of strain energy stored during the blunting process of the prefatigued crack provides, along with the rapid loading rate imposed by the pressure bars, highest dynamic loading conditions early in the crack growth process. Therefore, higher crack velocities are expected just after crack initiation. Crack arrest occurred under a fibrous failure mode at about 15 mm from the prefatigued crack tip.

By continuing the analysis beyond the initiation time, the dynamic fracture toughness of a rapidly propagating "fibrous" crack can be obtained. This requires iterative analyses to be performed until the experimental and analytical crack-opening displacements (COD_X) are matched. New experiments need to be performed with this technique to measure a crack-arrest toughness. This involves the use of a finer crack gage technique to identify the precise time of arrest. The use of pressure bars of different lengths may also be used to obtain controlled crack-arrest lengths. In parallel, a fracture criterion for ductile crack growth, independent of specimen geometry, needs to be developed. Two potential candidates are the two crack-tip integrals previously mentioned, J' and T^*.

Figure 8 compares the quasi-static fracture toughness [*21*] to the dynamic fracture-initiation toughness obtained with the coupled pressure bars. With increased loading rate, the transition temperature is increased by about 25°C, whereas the upper-shelf fracture toughness is increased by at least 70%. The increase of the transition temperature and upper-shelf toughness with loading rate seems to be typical of ductile steels exhibiting a strong strain-rate dependence [*20*].

Dynamic fracture-initiation toughness and crack-arrest toughness obtained from the series of wide-plate crack arrest experiments, WP-1, are compared in Fig. 9. The dynamic initiation toughness is found to be similar to the crack-arrest toughness in the transition regime (37 to 50°C). At temperatures above 50°C, the dynamic initiation toughness rises with the crack-arrest toughness. The dynamic initiation toughness appears to provide an estimate of the crack-arrest toughness, at least up to 75°C.

Table 2 summarizes the failure modes involved in crack initiation, propagation, and arrest for the specimens tested over the temperature range 37 to 100°C. Such observations match those made with the WP-1 series [*1*], in that cleavage fracture was observed at temperatures lower than

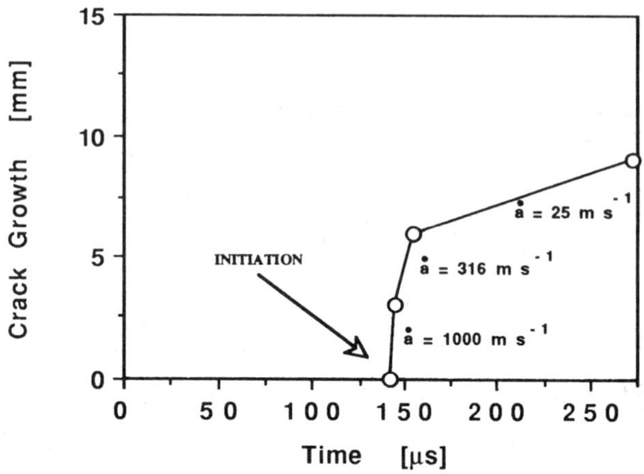

FIG. 7—*Crack growth history of the A533B specimen tested at 100°C.*

92°C. More dynamic initiation and crack-arrest experiments need to be performed at elevated temperatures to establish if this correspondence prevails over the upper-shelf regime. Along with this investigation, the eventual K_{Id}–K_{Ia} relationship based on the argument that the inability of an arrested crack to reinitiate is related to the properties of a stationary crack being loaded dynamically, remains to be established. The former need to be demonstrated not only for arresting cracks preceded by a rapid cleavage fracture, but also by a rapid fibrous fracture [*22*].

The CPB technique has been demonstrated to promote rapid crack propagation successfully in small A533-B specimens at a temperature never reached with other experimental techniques. Based on an observed leveling of the upper-shelf dynamic initiation toughness of certain low-strength steels [*20*], we anticipate that the technique can be used to evaluate the dynamic

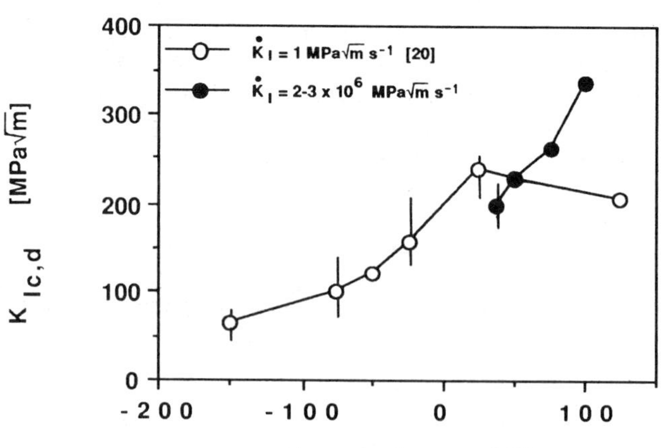

FIG. 8—*Comparison of A533B quasi-static and dynamic fracture initiation toughness over the temperature range 37 to 100°C.*

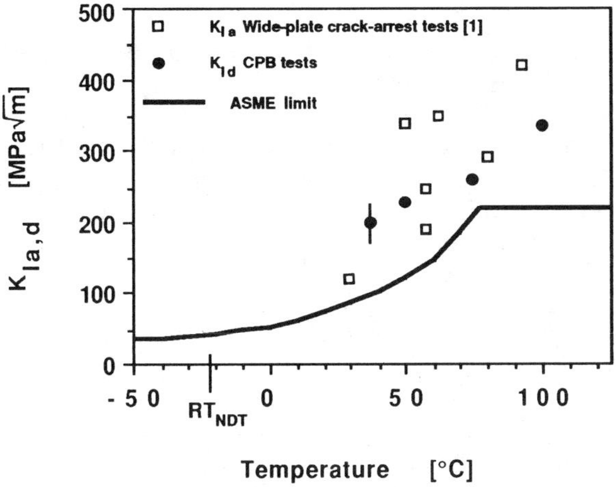

FIG. 9—*Dynamic fracture-initiation toughness* (K_{Id}) *and crack-arrest toughness* (K_{Ia}) *of A533B steel. The fracture-toughness values prescribed by Section XI of the ASME Boiler and Pressure Vessel Code are indicated.*

fracture properties of A533B steel up to 320°C, the service temperature of nuclear power plants. It is estimated that with a specimen of crack length, thickness, and remaining ligament size of at least 70 mm, valid dynamic fracture initiation toughness of A533B steel, based on Relation 3, can be evaluated from 100 to 320°C. For a lower bound of the dynamic yield stress of 450 MPa, a valid initiation fracture toughness of 500 MPa · $m^{1/2}$ can be measured.

The CPB technique presents other unique features with regard to the characterization of dynamic fracture properties. The technique has the potential for extracting dynamic initiation, propagation, and arrest toughnesses from the analysis of a single specimen with a dynamic viscoplastic fracture simulation code. The analysis is optimum because the specimen size is minimized with regard to the measure of a valid plane-strain fracture toughness (large specimens with finely discretized meshes become intractable for extended fracture simulations). This technology permits the investigation of the correlation between dynamic initiation toughness and crack-arrest toughness, as well as the relation between dynamic propagation toughness and crack velocity for a fibrous type of fracture. Furthermore, because energy storage is independent of specimen size, the possible influence of specimen size on dynamic fracture toughness can be examined systematically.

TABLE 2—*Summary of the failure modes involved in the CPB specimens.*

Testing Temperature, °C	Initiation: Fibrous Mode, %	Propagation: Failure Mode as a Function of Crack Extension a-a_o	Arrest: Fibrous Mode, %
37	30	cleavage for a-a_o varying from 0.5 to 12.5 mm	100
50	85	cleavage for a-a_o varying from 0.5 to 12.0 mm	100
75	100	cleavage for a-a_o varying from 1 to 9.5 mm	100
100	100	no cleavage observed (total growth 15 mm)	100

Conclusions

A novel dynamic fracture experiment using coupled pressure bars (CPB) has been developed to characterize the dynamic fracture properties of high-toughness materials using small specimens. The dynamic upper-shelf regime of a nuclear pressure vessel, A533B steel, was investigated with the CPB technique in the temperature range from 37 to 100°C (60 to 123°C above the nil-ductility transition temperature). The dynamic fracture-initiation toughness was deduced from numerical simulations of the experiment using a dynamic viscoplastic finite element code. The following conclusions can be drawn from this work:

1. With an increased loading rate from 1 (quasi-static) to 2×10^6 MPa · $m^{1/2}$ s^{-1}, the transition temperature is increased by about 25°C, whereas the upper-shelf initiation fracture toughness is increased by 70%.
2. The dynamic initiation toughness seems to provide an estimate of the crack-arrest toughness, at least up to 75°C. However, the theoretical basis for this remains to be established.
3. At 100°C, the rapid crack propagation event occurred with a fully fibrous type of fracture at velocities reaching 1000 m s^{-1}. The fracture processes involved during the propagation and arrest events for temperature ranging from 37 to 100°C were found to agree with wide-plate crack-arrest results.
4. The CPB technique has been proven to measure initiation toughness with small specimens up to 337 MPa · $m^{1/2}$. The minimum specimen size was found to be dictated by the size criterion required to measure a valid plane-strain fracture toughness and not by the ability of the CPB technique to promote rapid crack initiation.
5. Furthermore, it appears that the CPB technique can also be extended to the measurement of propagation and arrest toughness for a wide range of failure modes (that is, cleavage to fibrous).

Acknowledgments

The authors express their thanks to Mr. V. Aaron for this technical support and to Dr. C. P. Leung for his participation in the analyses. This work was carried out under Electric Power Research Institute (EPRI) Program RP-2426-22. The EPRI program manager is Dr. R. Viswanathan.

References

[1] Bass, R. B., Pugh, C. E., Merkle, J. G., Naus, D. J., and Keeney-Walker, J., in *Fracture Mechanics: Nineteenth Symposium, ASTM STP 969*, T. A. Cruse, Ed., American Society for Testing and Materials, Philadelphia, 1988, p. 691.

[2] Hahn, G. T., Hoagland, R. G., Lereim, J., Markworth, A. J., and Rosenfield, A. R., in *Crack Arrest Methodology and Applications, ASTM STP 711*, American Society for Testing and Materials, Philadelphia, 1980, p. 289.

[3] Rosenfield, A. R., Mincer, P. N., and Marschall, C. W., in *Fracture Mechanics: Eighteenth Symposium*, ASTM STP 945, American Society for Testing and Materials, Philadelphia, 1988, p. 73.

[4] Kussmaul, K. and Gillot, R., *Journal of Pressure Vessel Technology*, Vol. 110, 1988, p. 129.

[5] Fourney, W. L., Rosenfield, A. R., Marschall, C. W., and Hoagland, R. G., *Metal Handbook*, 9th ed., Vol. 8, 1985, p. 288.

[6] Marschall, C. W., Mincer, P. N., and Rosenfield, A. R., in *Fracture Mechanics: Fourteenth Symposium, ASTM STP 791*, Vol. 2, American Society for Testing and Materials, Philadelphia, 1983, p. 295.

[7] Crosley, P. B., et al., Report NUREG/CR-3261, University of Maryland, College Park, MD, 1983.

[8] Couque, H., Hudak, S. J., Jr., and Lindholm, U. S., *Journal de Physique*, Suplément No. 9, Tome 49, 1988, p. 347.

[9] Couque, H., Hudak, S. J., Jr., and Aaron, V., to appear in *Journal of Experimental Mechanics*.

[10] Dexter, R. J., Hudak, S. J., Jr., Reed, K. W., Polch, E. Z., and Kanninen, M. F., in *Numerical Methods in Fracture Mechanics*, A. R. Luxmore, et al., Eds., Pineridge Press, Swansea, U.K., 1987, p. 173.
[11] Dexter, R. J. and Chan, K. S., *Journal of Pressure Vessel Technology*, Vol. 112, 1990, p. 218.
[12] Kostrov, B. V. and Nikitin, L. V., *Archiwum Mechaniki Stovowanej*, Vol. 22, 1970, p. 749.
[13] Freund, L. B., *Journal of Elasticity*, Vol. 2, 1972, p. 341.
[14] Atluri, S. N., *Engineering Fracture Mechanics*, Vol. 16, 1982, p. 341.
[15] Paris, P. C., in written discussion to Landes, J. D. and Begley, J. A., in *Fracture Toughness (Proceedings of the 1971 National Symposium on Fracture Mechanics) Part II, ASTM STP 514*, American Society for Testing and Materials, Philadelphia, 1972, p. 1.
[16] Landes, J. D. and Begley, J. A., in *Developments in Fracture Mechanics Test Methods Standardization, ASTM STP 632*, American Society for Testing and Materials, Philadelphia, 1977, p. 57.
[17] Hutchinson, J. W., *Journal of Applied Mechanics*, Vol. 50, 1983, p. 1042.
[18] Moran, B., Asaro, R. J., and Shih, C. F., *Metallurgica Transaction A*, Vol. 22A, 1991, p. 161.
[19] Costin, L. S., Duffy, J., and Freund, L. B., in *Fast Fracture and Crack Arrest, ASTM STP 627*, American Society for Testing and Materials, Philadelphia, 1976, p. 301.
[20] Couque, H., Asaro, R. J., Duffy, J., and Lee, S., *Metallurgica Transaction A*, Vol. 19A, 1988, p. 2179.
[21] Edmonds, D. P. and McGowan, J. J., Report NUREG/CR-3744, Vol. 2, Oak Ridge National Laboratory, Oak Ridge, TN, 1984, p. 96.
[22] Irwin, G. R., in *Fast Fracture and Crack Arrest, ASTM STP 627*, American Society for Testing and Materials, Philadelphia, 1983, p. 7.

DISCUSSION

M. T. Miglin[1] (*written discussion*)—Do you observe any ringing, that is, load oscillations, in your load versus time traces?

H. Couque (*author's closure*)—As mentioned in the second paragraph of the Results and Analysis section, the specimen-loading history was estimated from strain measurements performed on the arms of a 4340 steel specimen tested with the CPB technique at a preload of 444 kN. No strain oscillations were recorded during the first 100 μs, corresponding to the failure duration of the starter specimen. During propagation, strain oscillations varying from 20 to 40% of the average amplitude were observed.

J. T. Merkle[2] (*written discussion*)—Has the sensitivity of crack driving force to near crack-tip finite-element mesh size when using viscoplastic analysis been examined? If so, what were the results? Oak Ridge National Laboratory (ORNL) has found that, for moving cracks, the crack driving force decreases as mesh size decreases when using viscoplastic analysis.

H. Couque (*author's closure*)—The effect of mesh refinement has been investigated using two-dimensional finite-element simulations of the A533B specimen tested at 100°C with two mesh sizes of linear dimension 2 and 3 mm. The computed integral J' and T^* were found to coincide up to the onset of crack initiation. Such results confirm similar investigation with regard to J' for the initiation event by Nakamura et al. [1]. During the propagation event, the dynamic propagation toughness was found to be mesh-dependent as in the mentioned work of Bass et al.

[1] Babcock & Wilcox, 1562 Beeson St., Alliance, OH 44601.
[2] Oak Ridge National Laboratory, P.O. Box 2009, Oak Ridge, TN. 37831-8049.

[2] reporting results from simulations of the wide-plate test WP-1.2. Such dependency may indicate the need to develop new fracture criteria for quasi-static and dynamic ductile crack growth.

References

[1] Nakamura, T., Shih, C. F., and Freund, L. B., *Engineering Fracture Mechanics*, Vol. 22, No. 3, 1985, p. 437.

[2] Bass, B. R., Keeney-Walker, J., and Schwartz, C. W., *Computers & Structures*, Vol. 32, No. 3, 1989, p. 815.

Hiroomi Homma,[1] *Yasuhiro Kanto,*[1] *and Kohji Tanaka*[1]

Cleavage Fracture Under Short Stress Pulse Loading at Low Temperature

REFERENCE: Homma, H., Kanto, Y., and Tanaka, K., **"Cleavage Fracture Under Short Stress Pulse Loading at Low Temperature,"** *Rapid Load Fracture Testing, ASTM STP 1130,* Ravinder Chona and William R. Corwin, Eds., American Society for Testing and Materials, Philadelphia, 1991, pp. 37–49.

ABSTRACT: Cracks in SM50A steel for welding structure, chilled to 233 and 193K, were loaded by various stress intensity pulses with durations of 20, 40, and 80 μs to generate experimental data of critical stress intensity levels for crack instability. Fracture surfaces were observed by a scanning electron microscope to examine cleavage nucleation origins ahead of the crack tips. The experimental results are discussed in terms of minimum time criterion and dislocation dynamics.

The critical stress intensity obtained by the minimum time criterion was higher at 233K than at 193K, whereas the minimum time was shorter at 233K. The cleavage nucleation origin approached the crack tip as the temperature decreased. Finally, the temperature dependence of the minimum time was interpreted by the dislocation dynamics.

KEYWORDS: crack instability, minimum time criterion, cleavage fracture, stress intensity, temperature

Crack instability is affected strongly by loading rate and test temperature, especially in body-centered cubic metal such as steel. The ASTM Test for Plane-Strain Fracture Toughness of Metallic Materials (E 399) is used for fracture toughness measurement under rapid loading. However, there is a restriction of loading rate so that a quasi-static stress field would develop in a specimen.

For rapid loading in which the stress field near a crack tip is affected significantly by the inertia, we have little reliable crack instability criteria and experimental data. The crack instability criteria proposed heretofore are classified as two types. One of them is based on the quasi-static concept that a crack becomes unstable immediately when a fracture parameter evaluated by full dynamic analysis equals or exceeds the critical value [*1,2*]. The other takes into account the time effect that the crack instability is brought about when the fracture parameter equals or exceeds the critical value for a certain period [*3–5*]. The period is called "minimum time" or "incubation time." Homma et al. [*6*] obtained minimum time values for several metals.

This work examines the minimum time or the incubation time for steel in which cleavage fracture is initiated at low temperature by short-pulse loading.

Experimental Procedure

Material and Specimens

Material used in the experiment was steel for welding structure, SM50A (Japanese Industrial Standards); its chemical composition and mechanical properties are indicated in Tables 1 and 2,

[1] Professor, associate professor, and graduate student, respectively, Toyohashi University of Technology, 1-1 Tempaku-cho, Toyohashi, Japan 441.

TABLE 1—*Chemical composition of SM50A.*

C Wt/%	Si Wt/%	Mn Wt/%	P Wt/%	S Wt/%	Fe Wt/%
<0.2	<0.55	<1.5	<0.04	<0.04	Remain

respectively. Three specimen configurations were prepared, as shown in Fig. 1. They had the same width and different length. A single-edged notch was machined in each specimen to initiate a 3- to 5-mm-long prefatigue crack from the notch root. When a short pulsative load was applied to the center of back surface of the specimen without any supports, the specimen was bent and swung back as a result of the inertia effect. This loading is called one-point bending [*7*]. A period of this movement depends on specimen dimensions. The specimen configurations shown in Fig. 1 were used to generate stress intensity histories with three different periods. The specimen dimensions were determined by the finite element method (FEM) parametric study in which the stress intensity time-histories were calculated for different lengths of the specimen, maintaining the width and the thickness constant. The calculation result indicated that 80-, 40-, and 20-μs stress intensity histories were generated in the 180-, 100-, and 50-mm-long specimens, respectively.

Test temperatures were 233 and 193K, which are lower than the transition temperature, 287K, of the steel [*8*] so that the whole fracture surface would be covered by cleavage.

Dynamic Loading Device and Measurement of Dynamic Stress Intensity Histories

The loading device is shown in Fig. 2. This consists of a launching device and a load transfer rod. A specimen was placed contacting with the round end of the load transfer rod. The launching device is made up of a cylindrical barrel, 2000 mm long by 20-mm inner diameter, a reservoir for nitrogen gas, and a solenoid valve. A projectile, 20 mm long by 20-mm diameter, was positioned at the end of the barrel, and the reservoir was filled with nitrogen gas of the desired pressure. The projectile was launched by opening the solenoid valve. The pressurized gas flew into the barrel at a high speed and pushed the projectile to accelerate it against the load transfer rod. Impact velocity was controlled by the gas pressure. In this experiment, the impact velocity was in the range of 10 to 50 m/s.

Dynamic stress intensity history experienced by a crack was measured by a strain gage mounted near the crack tip according to the method developed by Dally and Sanford [*9*]. The position of the strain gage is shown in Fig. 1. If the axis of the strain gage is selected as shown in Fig. 1, strain measured by the gage is related to the term of $r^{-1/2}$ alone, even though strain field near the crack tip is expressed by a series representation including three terms of $r^{-1/2}$, r^0, and $r^{1/2}$, where r is distance from the crack tip; an example of output of the strain gage for the 50-mm-long specimen is shown in Fig. 3. The duration at the half amplitude of the first pulse is about 20 μs, and the amplitude of the first pulse is twice as large as those of the subsequent pulses. The crack instability can take place by the first pulse if the amplitude is large enough.

TABLE 2—*Mechanical properties of SM50A.*

Yield Point, MPa	Tensile Strength, MPa	Elongation, %	Strain h · e, N	Young's Modulus, MPa
422	588	31	0.25	206 000

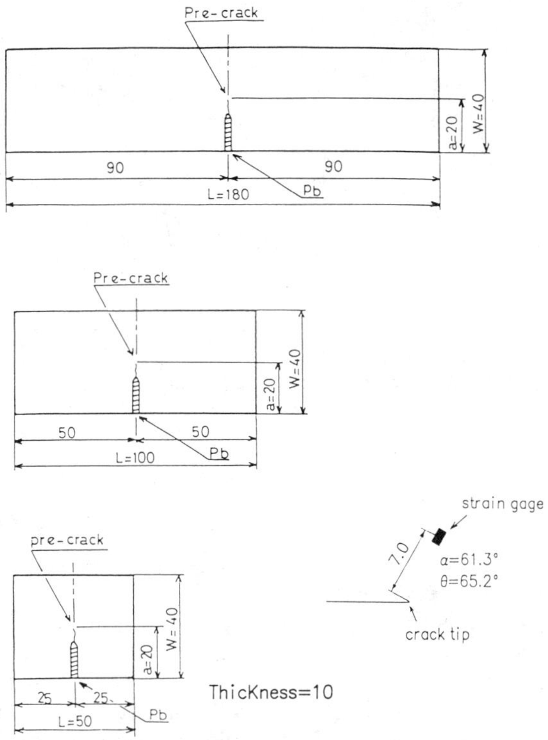

FIG. 1—*Specimen configuration and strain gage location (units in millimetres).*

Results

Critical Stress Intensity for Crack Growth

For each test temperature and each specimen configuration, four or five specimens having a crack of identical length were prepared and a stress intensity pulse of a different amplitude was

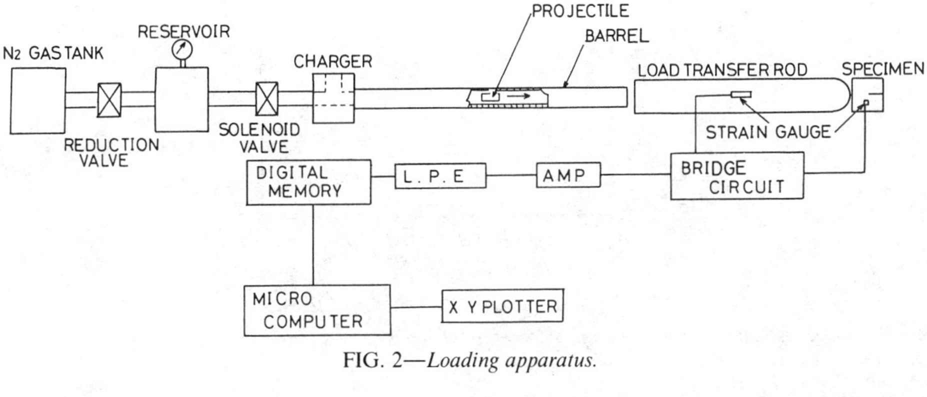

FIG. 2—*Loading apparatus.*

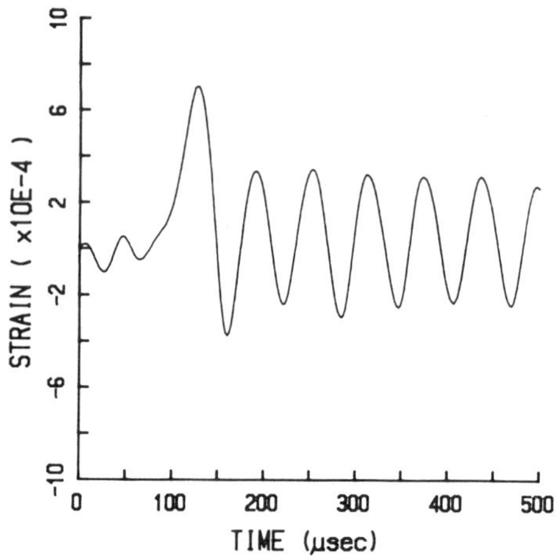

FIG. 3—*Output of the strain gage near the crack tip (L = 50 mm).*

applied to each specimen. After the impact, each specimen was examined using an optical microscope to see whether the crack had grown.

The maximum value of a dynamic stress intensity pulse measured by a strain gage mounted near a crack tip is plotted against the pulse duration as a parameter of the crack growth event in Figs. 4 and 5. The open circle means that the crack did not grow under the stress intensity pulse, and the solid circle means that the crack grew under the pulse. A critical value of dynamic stress intensity for the crack initiation was defined as the middle of the adjacent solid and open circles.

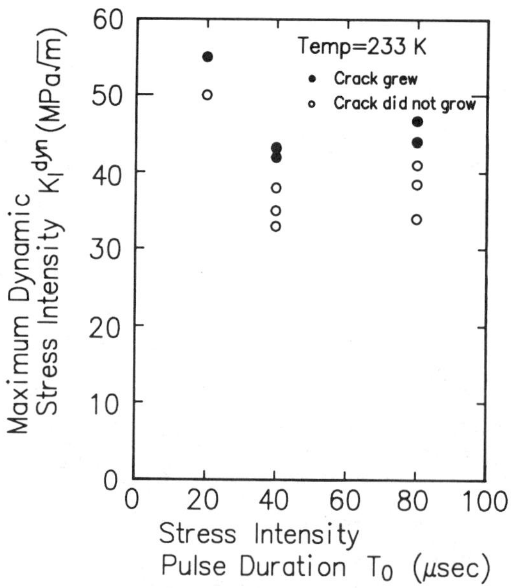

FIG. 4—*Maximum stress intensity as a function of stress intensity pulse duration for a crack growth event at 233K.*

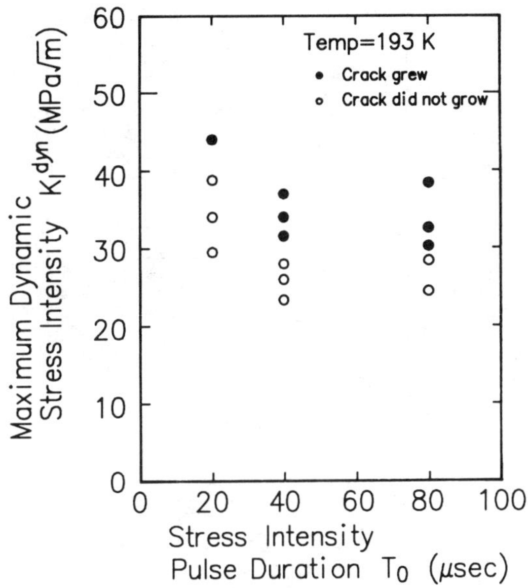

FIG. 5—*Maximum stress intensity as a function of stress intensity pulse duration for a crack growth event at 193K.*

The critical value is shown as a function of the stress intensity pulse duration in Fig. 6. The critical values for the 40- and 80-μs durations are almost the same and those are much smaller than the value for the 20-μs duration. It is also seen that there is temperature dependence of the critical value.

Observation of Cleavage Fracture

Fracture surface near the initial crack tip was observed by an scanning electron scope. Microscopic fractographs are shown in Figs. 7 and 8. For all the experimental conditions, the whole of fracture surface is covered by the cleavage.

The cleavage is always nucleated ahead of the initial crack tip. The nucleation origin can be detected from flow directions of river patterns. First, a chevron pattern was found on a photograph of low magnified fracture surface. Its sharp edge indicates a macroscopic fracture origin. Then, several fractographs near the origin were taken with high magnification (×400). Observing the flow directions of the river patterns, all the rivers were traced to their original source. The source was defined as the nucleation origin of the cleavage. Because only one chevron pattern was observed on each specimen, a single cleavage origin was found for each specimen.

The distance between the initial crack tip and the nucleation origin was measured and is shown as a function of the duration of the stress intensity pulse in Fig. 9. The nucleation origin is almost independent of the pulse duration, but significantly dependent on the test temperature. The nucleation distance is shown as a function of the test temperature in Fig. 10, comparing it with the result for pressure vessel steel A508S [*10*], which has similar mechanical properties and chemical composition to those of this steel. Hou's et al. [*11*] experimental results show that the nucleation distance for another steel decreased as the temperature was lowered and that the cleavage nucleation origin corresponded to the site at which the principal stress reached the maximum ahead of the crack tip blunted by the plastic deformation.

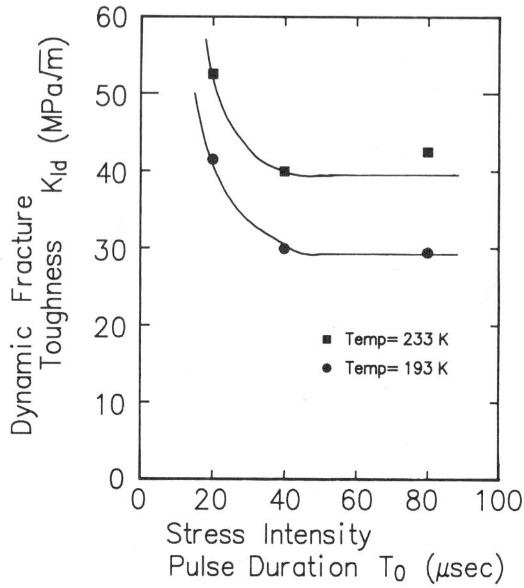

FIG. 6—*Maximum stress intensity for crack initiation as a function of pulse duration.*

Discussion

The critical maximum value of dynamic stress intensity pulse steeply increased at the duration of 20 μs, as shown in Fig. 6. The steep increase is contrary to the effect of loading rate on the dynamic fracture toughness that the higher loading rate makes yield strength of the material at the crack tip increase to lower the dynamic fracture toughness because the loading rate is higher in the stress intensity pulse with the 20-μs duration than in those with 40- and 80-μs durations.

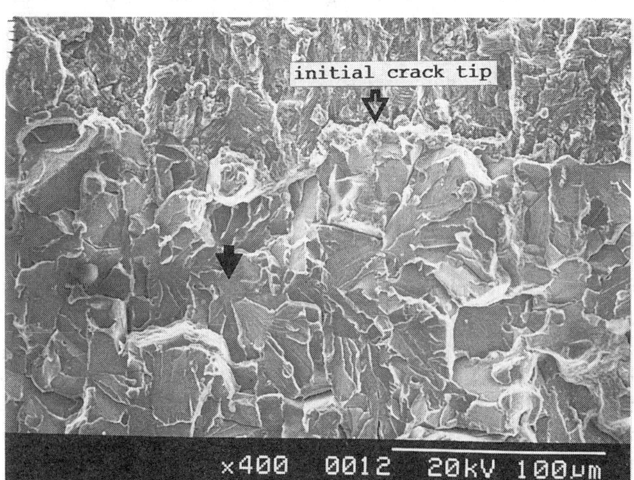

FIG. 7—*Microscopic fractograph indicating cleavage fracture from 20-μs duration at 193K.*

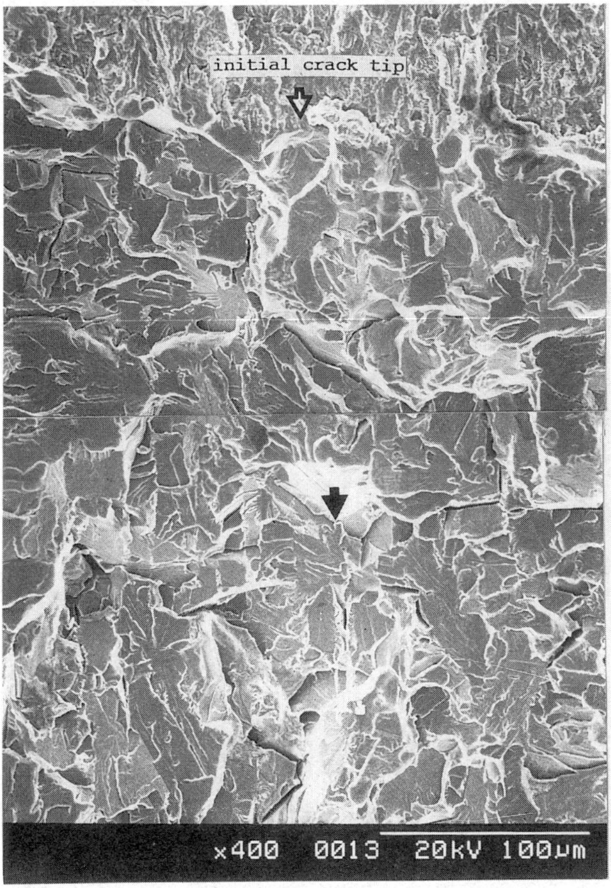

FIG. 8—*Microscopic fractograph indicating cleavage fracture from 80-μs duration at 233K.*

This can be explained in terms of the minimum time or the incubation time criterion [*6*]. According to the minimum time criterion, a crack can be initiated if the stress intensity at the crack tip equals or exceeds the critical value for a certain period. The period is called the minimum time. When this criterion was used, the critical value for crack initiation was considered to be constant because the loading rate was very fast and changed within the range of less than one order among the three pulse durations. The minimum time was estimated by the method used in the previous work [*6*]. At first, a certain time was taken as a trial minimum time and the critical stress intensity value was determined for the three durations according to the minimum time criterion. If the critical stress intensity value for the shortest duration was greater than the values for the other durations, the trial minimum time was increased for the second trial. This procedure was continued until the difference among the critical values for the three pulse durations was minimized. The obtained minimum times are 16 and 22 μs for 233 and 193K, respectively. The critical stress intensity values for the minimum times are shown in Fig. 11. Average values of them are 34.2 and 25.7 MPa $\cdot$ $m^{1/2}$ for those temperatures, respectively.

The nucleation origin of the cleavage is considered in association with the plastic zone. Dynamic yield strength of the material was estimated by the equation given in ASTM Method E

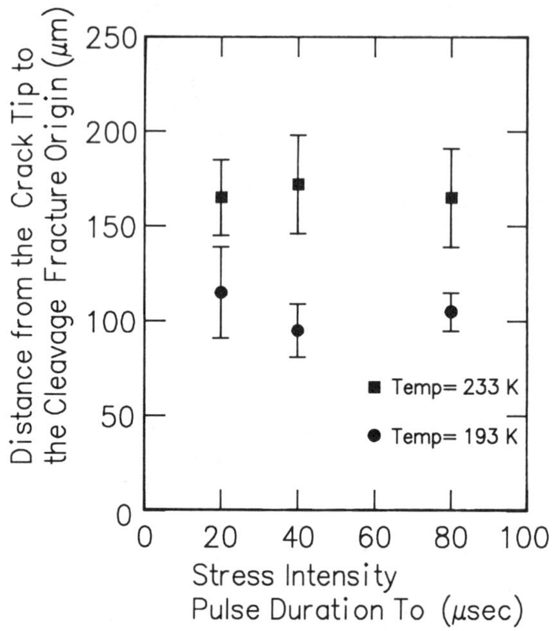

FIG. 9—*Distance from initial crack tip to cleavage fracture origin.*

399, Annex 7. The dynamic yield strength was 694 MPa for 233K and 80-μs fracture time and 792 MPa for 193K and 80-μs fracture time. The small scale yielding and plane strain conditions prevail near the initial crack tip at the crack initiation. The plastic zone size was calculated by the following equation:

$$r_p = (1/6\pi)(K/\sigma_{yd})^2 \quad (1)$$

The calculated results were 129 and 56 μm for 233 and 193K, respectively. Comparing these results with the nucleation origin of the cleavage shown in Fig. 10, it is seen that the cleavage fracture was initiated a distance one to two times the plastic zone size from the initial crack tip. However, because average grain diameter of the used steel is around 50 μm, only one or two grains exist in the plastic zone. Therefore, the plastic zone size given by Eq 1 based on continuum mechanics may not correspond to the real one. The cleavage was nucleated two or three grains ahead of the crack tip. When the crack is loaded by the critical stress intensity, the elastic stress at the cleavage origin can be calculated as

$$\begin{aligned}\sigma &= K/\sqrt{2\pi r} \\ &= 25.7/\sqrt{2\pi \times 100 \times 10^{-6}} = 1.03 \text{ GPa for 193K} \\ &= 34.2/\sqrt{2\pi \times 165 \times 10^{-6}} = 1.06 \text{ GPa for 233K}\end{aligned}$$

The elastic stress is almost the same for both the test temperatures.

The minimum time or the incubation time obtained by the experiment is discussed in terms of dislocation dynamics. Hahn et al. [*12*] reviewed the proposed dislocation models for the cleavage fracture and explained their experimental data using a dislocation model. In all the disloca-

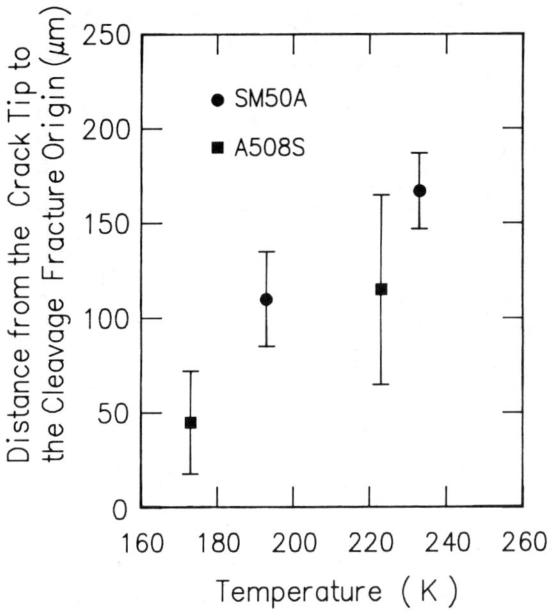

FIG. 10—*Cleavage fracture origin as a function of temperature.*

tion models, the cleavage fracture is initiated by the pile-up of dislocations against an obstacle, such as the grain boundary.

According to the dislocation dynamics, the dislocation velocity v is given by

$$v = v^* \exp\{-D/\sigma\} \tag{2}$$

where

v^* = the terminal velocity,
D = the drag stress, and
σ = the applied stress.

When a crack is loaded by the critical stress intensity, the dislocation velocity at the position 50 μm from the crack tip is estimated by the above equation of the yield strength and applied stress at the position are substituted into D and σ, respectively, as follows:

$$\begin{aligned} v(193) &= v^* \exp\{-792/\sigma_1\} \\ v(233) &= v^* \exp\{-694/\sigma_2\} \end{aligned} \tag{3}$$

where σ_1 and σ_2 are the applied stresses at the position 50 μm from the crack tip for each temperature and $\sigma_1 = 25.4/\sqrt{2\pi \times 50 \times 10^{-6}}$ and $\sigma_2 = 34.2/\sqrt{2\pi \times 50 \times 10^{-6}}$. Then, the ratio of $v(193)$ to $v(233)$ is

$$\frac{v(193)}{v(233)} = 0.74 \tag{4}$$

This suggests that because the dislocation velocity is smaller at 193K than at 233K, it takes

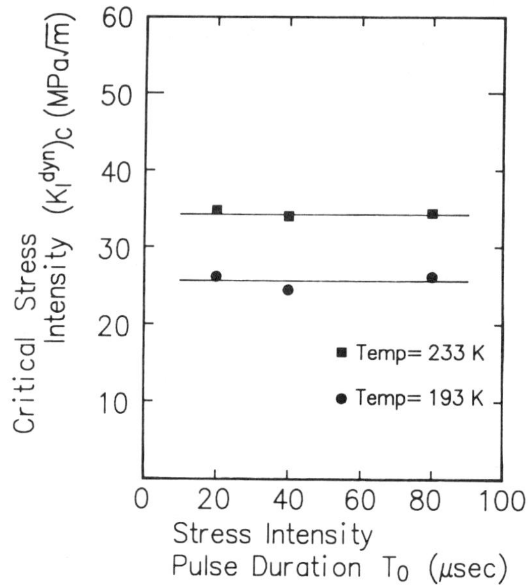

FIG. 11—*Critical stress intensity for crack initiation based on the minimum time criterion.*

longer for the dislocation to travel the same distance at 193K. One of the reasons for the temperature dependence of the minimum time may be attributed to the dislocation velocity.

On the other hand, emission of dislocations from a loaded crack tip was analyzed by Yokobori et al. [*13*]. They assumed that the dislocation movement is the thermally activated process. According to their results, dynamic behavior of a dislocation in a group of dislocations was similar to that of an isolated dislocation. The total number of dislocations emitted from the source was given for pure iron as

$$N = 2.637\ Q^4 \tag{5}$$

where

$$Q = \left(\frac{\tau}{\tau_0}\right)^{0.8}\left(\frac{t}{t_0}\right) \tag{6}$$

and

τ = applied stress rate to a dislocation,
τ_0 = nominal applied stress rate to an isolated dislocation,
t = time, and
t_0 = the time necesssary for an isolated dislocation to travel a specific distance, l.

t_0 is expressed by

$$t_0 = (4l)^{1/4}\left(\frac{\tau^*}{\tau_0}\right)^{3/4} \tag{7}$$

where τ^* is shear stress necessary to move an isolated dislocation at a velocity of 1 cm/s.

When the emission source of dislocations is a crack tip, τ/τ_0 corresponds to the stress concentration factor as a result of the crack. For the same crack length and the same loading configuration, this would be constant. If τ^* is proportional to macroscopic yield strength of the material, the number of emitted dislocations from the crack tip loaded by the critical stress intensity is given as

$$N = \beta t^{4} \left(\frac{\tau_0}{\sigma_{ys}}\right)^3 \tag{8}$$

where β is a constant. The stress rate is directly related to the increasing rate of stress intensity, K, in this experiment.

The ideal cleavage fracture stress, σ_c, is related to the mechanical properties of the material as $\sigma_c = 2\tau_c = 4\{E\gamma/\pi(1-\nu^2)d\}^{1/2}$, where E is Young's modulus, γ is surface energy, ν is Poisson's ratio, and d is grain size [*14*]. The parameters in the above equation are essentially independent of the temperature. At the cleavage nucleation origins, the normal stresses are identical for both temperatures, as indicated previously. It may be considered that pile-up of the same number of dislocations is necessary to nucleate the cleavage for both temperatures. The time required for emission of the same number of dislocations is calculated by Eq 8 for both temperatures as

$$\frac{t(193)}{t(233)} = \left[\frac{\tau^*(193)}{\tau^*(233)}\,\frac{\tau_0(233)}{\tau_0(193)}\right]^{3/4} \tag{9}$$

Substitution of the yield strength and the increasing rate of stress intensity into τ^* and τ_0 gives the ratio of the time as

$$\frac{t(193)}{t(233)} = 1.37 \tag{10}$$

This ratio compares well with the ratio of the minimum times for both the temperatures, 1.38. From this result, it can be deduced that the minimum time obtained by the experiment would be regarded as the time required to generate the number of dislocations that is large enough to nucleate the cleavage fracture in the vicinity of an obstacle for their pile-up.

Conclusions

Experiments on crack instability under short pulse loading were made at low temperatures of 233 and 193K using steel for welding structure SM50A so that cleavage fracture would predominantly take place at the crack initiation. The following conclusions were obtained:

1. The maximum stress intensity experienced by a crack at the crack instability was much larger for 20-μs-duration stress intensity pulse than for 40- and 80-μs-duration stress intensity pulses. This was well explained by the minimum time criterion.
2. The obtained minimum times were 16 and 22 μs for the test temperatures of 233 and 193K, respectively.
3. The cleavage fracture nucleation origins were around 100 and 165 μm from the initial crack for 233 and 193K, respectively.
4. The minimum time was interpreted as the time required to generate and pile up the number of the dislocations that is large enough to bring about the cleavage fracture ahead of the obstacle.

References

[1] Achenbach, J. D. and Brock, L. M., in *Dynamic Crack Propagation*, G. C. Sih, Ed., Noordhoff, The Netherlands, 1973, p. 529.
[2] Lehnigk, S. H., in *Dynamic Crack Propagation*, G. C. Sih, Ed., Noordhoff, The Netherlands, 1973, p. 333.
[3] Kalthoff, J. F. and Shockey, D. A., *Journal of Applied Physics*, Vol. 43, No. 3, 1977, p. 986.
[4] Homma, H., Shockey, D. A., and Murayama, Y., *Journal of the Mechanics and Physics of Solids*, Vol. 31, No. 3, 1983, p. 261.
[5] Kalthoff, J. F., *Engineering Fracture Mechanics*, Vol. 23, No. 1, 1986, p. 289.
[6] Homma, H., Shockey, D. A., and Hada, S., in *Fracture Mechanics: Seventeenth Volume, ASTM STP 905*, American Society for Testing and Materials, Philadelphia, 1986, p. 683.
[7] Bohme, W. and Kalthoff, J. F., *International Journal of Fracture*, Vol. 20, No. 1, 1982, p. 139.
[8] Fujimoto, S., Aoki, H., Nakagomi, T., Terada, T., and Wakiyama, K., *Journal of Structure and Construction Engineering*, Architectural Institute of Japan, No. 415, 1990, p. 89.
[9] Dally, J. W. and Sanford, R. J., *Experimental Mechanics*, Vol. 27, No. 4, 1987, p. 381.
[10] Tanaka, Y., "A Study of Evaluation of Integrity of Primary System of Light Water Reactor Using Fracture Mechanics," Report of Japan Welding Engineering Society, JWES-AE-8701, Tokyo, 1987, p. 52.
[11] Hou, C. X., Cai, Q. G., and Su, Y., in *Advances in Fracture Research, Proceedings of ICF7*, K. Salama et al., Eds., Pergamon Press, Houston, 1989, p. 2501.
[12] Hahn, G. T., Averbach, B. L., Owen, W. S., and Cohen, M., in *Fracture*, B. L. Averbach et al., Eds., MIT Press, Swampscott, MA, 1959, p. 91.
[13] Yokobori, T., Jr., Yokobori, T., and Kamei, A., "Computer Simulation of Dynamics of Dislocation Group under Constant Applied Stress Rate," *Transaction of Japan Society for Mechanical Engineers*, Vol. 42, No. 358, 1976, p. 1652.
[14] Smith, E., in *Proceedings of Conference on Physical Basis of Yield and Fracture*, Oxford, Institute of Physics and Physics Society, London, 1966, p. 36.

DISCUSSION

Question 1

A. R. Rosenfeld[1] (*written discussion*)—Have you done any experiments in which stable crack growth occurred before cleavage?

H. Homma (*author's closure*)—No, I have not. In this work, our attention was placed on crack initiation completely brought about by cleavage. So, the test temperature was lower than the transition temperature.

Question 2

A. R. Rosenfeld (*written discussion*)—Does the analysis you describe only pertain to cleavage without stable growth?

H. Homma (*author's closure*)—In the analysis, it is considered that the emanation of dislocations blunts the crack tip, but it is not considered that stable crack growth takes place during the emanation of dislocations.

Question 3

A. R. Rosenfeld (*written discussion*)—Will you investigate stable growth before cleavage in the future?

[1] Battelle—Columbus, Columbus, Ohio.

H. Homma (*author's closure*)—Cleavage fracture after stable crack growth is a very interesting problem. However, because it is a quite complicated phenomenon, I do not have any research plan concerning this problem.

J. D. Landes[2] (*written discussion*)—Some people measure distance ahead of the crack tip relative to the crack tip opening displacement (CTOD). Do you know the CTOD values at fracture (S_c) so that the distance to the initiation sites can be measured relative to them?

H. Homma (*author's closure*)—Although we did not measure CTOD, the stretched zone width (*szw*) was measured by observation of the fracture surface with a scanning electron microscope. It was around 10 to 15 μm for 233K and 7 to 10 μm for 193K. Therefore, the cleavage fracture was initiated 10 to 17 times the *szw* from the initial crack tip.

[2] University of Tennessee, Knoxville, Tennessee.

Mark T. Kirk,[1] *Joseph P. Waskey,*[1] *and Robert H. Dodds, Jr.*[2]

A Procedure for Drop-Tower Testing of Shallow-Cracked, Single-Edge Notched Bend Specimens

REFERENCE: Kirk, M. T., Waskey, J. P., and Dodds, R. H., Jr., **"A Procedure for Drop-Tower Testing of Shallow-Cracked, Single-Edge Notched Bend Specimens,"** *Rapid Load Fracture Testing, ASTM STP 1130,* Ravinder Chona and William R. Corwin, Eds., American Society for Testing and Materials, Philadelphia, 1992, pp. 50–75.

ABSTRACT: Unacceptably pessimistic predictions of structural reserve capacity often result from the use of conventional initiation fracture-toughness values (for example, J_{Ic}, δ_c) measured using deeply cracked bend specimens. Conversely, favorable comparison between predictions and structural fracture behavior have been reported when the predictions are based on toughness values measured in such a manner that the crack-tip constraint and the loading rate closely match those experienced in service. As a result, considerable attention has been paid recently to the elevated fracture initiation resistance of cracks shallower than the 0.45 to 0.75 crack length/width (a/W) used in standardized tests. However, standardized procedures for estimating the fracture initiation resistance of shallow cracks, particularly at the high loading rates characteristic of certain severe service conditions, have yet to be developed. In this investigation, techniques useful for estimating the load, load-line displacement, and the time of crack initiation during impact tests of single-edge notched bend specimens having shallow fatigue cracks ($a/W \approx 0.1$) were developed and validated. The bending stress distribution across the specimen midway between the support and loading points, determined based on strain measurements and a uniaxial stress-strain relationship, was used to estimate the load imparted to the specimen by the impactor. Load-line displacement was estimated using four noncontacting transducers positioned along the underside of the bend specimen, whereas the time of crack initiation was inferred from strain readings taken from the elastically loaded region behind the fatigue crack tip. The applicability of these procedures was demonstrated at an impact loading rate of 4.88 m/s, whereas control experiments at quasi-static loading rates were used to demonstrate the accuracy of these new procedures for both low- and high-strain-hardening steel alloys.

KEYWORDS: shallow cracks, impact testing, *J*-integral, initiation fracture toughness, experimental procedures

The deep cracks used in standard fracture-toughness test specimens (initial crack length/specimen width, a_o/W, between 0.45 and 0.75) maximize the crack-tip stress triaxiality, thereby ensuring measurement of a lower-bound, geometry-independent fracture-toughness value. However, use of such lower-bound values in structural fracture integrity assessments often leads to unacceptably pessimistic predictions of structural reserve capacity [*1,2*] because of a mismatch between the degree of triaxial plane strain constraint in the test specimen and in the structure. Accordingly, considerable attention has been paid to the elevated fracture initiation resistance characteristic of cracks shallower than the 0.45 to 0.75 a/W used in standard test specimens

[1] Mechanical engineer and mechanical engineering technician, respectively, Fatigue and Fracture Branch of the David Taylor Research Center, Annapolis, MD 21402.

[2] Professor, Department of Civil Engineering, University of Illinois, 3129-D Newmark Lab., 205 N. Matthews Ave., Urbana, IL 61801.

[3–9]. Although it is commonly recognized that a geometry-independent toughness value has been sacrificed in the process, the improved accuracy of structural failure predictions that can be achieved using toughness values measured with specimens having shallow cracks [1,2] provides a useful engineering tool.

The preceding comments suggest that the accuracy of a structural fracture integrity assessment depends on matching the crack depth in the fracture-toughness specimen to the crack depth in the structure. For ferritic materials, there is an additional requirement that the loading rate applied to the fracture specimen be similar to that experienced in service as a result of the dependence of fracture toughness, fracture mode, and flow properties on the rate of load application for these materials. Accordingly, this paper presents a procedure for determining the initiation fracture toughness of impact-loaded ferritic materials having shallow initial crack depths ($a_o/W \approx 0.1$). The scope of this procedure is limited to loading rates below those at which the equations used to interpret the experimental data need to account for kinetic energy terms.

The value of the J-integral at crack initiation in a shallow-cracked bend specimen can be estimated for plane strain conditions by the following equation proposed by Sumpter [8]:

$$J_i = \frac{K^2 \cdot (1 - \nu^2)}{E} + \frac{\eta_p}{B \cdot b} \cdot \int_0^{\Delta_{pLL}{}^i} P d\Delta_{pLL} \tag{1}$$

where

K = linear elastic stress intensity factor calculated from the applied load at crack initiation,
ν = Poisson's ratio,
E = Young's modulus,
$\eta_p = 0.32 + 12.0(a/W) - 49.5(a/W)^2 + 99.8(a/W)^3$ for $a/W < 0.282$,
$= 2.0$ for $a/W \geq 0.282$,
Δ_{pLL} = plastic component of load line displacement,
P = load,
$\Delta_{pLL}{}^i$ = plastic component of load line displacement at crack initiation,
B = specimen thickness, and
b = specimen remaining ligament.

This equation indicates that measurements of load, load-line displacement, and detection of crack initiation are required to evaluate J_i. In this investigation, procedures useful for measuring these quantities during impact experiments on SE(B) specimens having shallow through cracks will be presented and validated.

Approach

To demonstrate the validity of candidate procedures for dynamic load, load-line displacement, and crack initiation measurements, two requirements must be fulfilled. First, the accuracy of such procedures must be demonstrated by comparison to control measurements. Second, it must be demonstrated that the transducers used to make these measurements have acceptable frequency response for the impact loading events of interest, and that kinetic energy terms can be ignored in the data analysis without undue error. To this end, the following steps were taken:

1. Quasi-static SE(B) experiments were performed so that the experimental load and load-line displacement values estimated using the candidate procedures for impact loading could be compared to control measurements made possible by the reduced loading rate.
2. Impact SE(B) experiments were performed to demonstrate that the data required by the candidate load and load-line displacement estimation procedures are obtainable and ana-

TABLE 1—*Tensile stress-strain properties of materials investigated.*[a]

Material	Loading Rate, mm/s	0.2% Offset Yield Strength, MPa	Ultimate Tensile Strength, MPa	Reduction in Area, %	Elongation over 25.4 mm Gage Length, %
Low hardening	0.01	745	841	66	. . .
		800	889	63	27
		793	883	62	29
	3454	827	986	61	31
		820	986	63	34
		841	. . .	61	30
High hardening (ASTM A515 Gr. 70)	0.01	287	543	52	34
		314	547	51	36
		297	536	51	32

[a] Tensile properties measured using a round bar specimen having an initial diameter of 7.95 mm and an initial gage length of 25.4 mm.

lyzable, and that the transducers used to make these measurements have adequate frequency response. Additionally, data from these experiments were used to demonstrate that an analysis procedure that ignores kinetic energy terms has acceptable accuracy.

3. A series of impact SE(B) experiments was performed, each specimen having experienced a different total load-line displacement so as to achieve different amounts of stable crack extension from the fatigue precrack. The results of these experiments were used to validate the candidate procedure for determining the time at which crack initiation occurs during impact loading.

Specimens for experiments performed as part of the first step were removed from both low-hardening (ultimate/yield [U/Y] ratio = 1.15) and high-hardening (U/Y ratio = 1.81) steel plate stock, whereas experiments for the second and third steps were performed using only the low-hardening material. The transverse tensile properties of these two materials at both quasi-static and impact rates are reported in Table 1. Square cross-section (50.8- by 50.8-mm) SE(B) specimens were used for all tests reported herein. The SE(B) specimens were taken in the T-S orientation; they had an initial a/W of approximately 0.1 and were tested with a span-to-width ratio of 4:1.

Experimental Procedures

SE(B) Testing

Quasi-static tests were performed in a screw-driven testing machine; displacement was applied to the specimen at a constant rate until the desired maximum displacement was achieved. Dynamic fracture tests were conducted in a drop tower with a load applied by a free-falling crosshead loaded with 669 kg of lead weights. The crosshead was dropped from 1.22 m to achieve a displacement rate of 4.88 m/s and arrested onto rigid stops after impacting and displacing the specimen by a prescribed amount. The SE(B) specimens were supported on two hardened anvils having the same width as the specimen. A pyramidal aluminum absorber (base of 25.4 by 38.1 mm, height at the apex of 25.4 mm, Brinell hardness of 22) was placed between the falling crosshead and the specimen. The impact flattened the absorber to a pancake approximately 2.5 mm thick. By flattening in this way, the absorber served as a deformable link between the specimen and the tup, allowing smooth load transfer to the specimen without the wild load

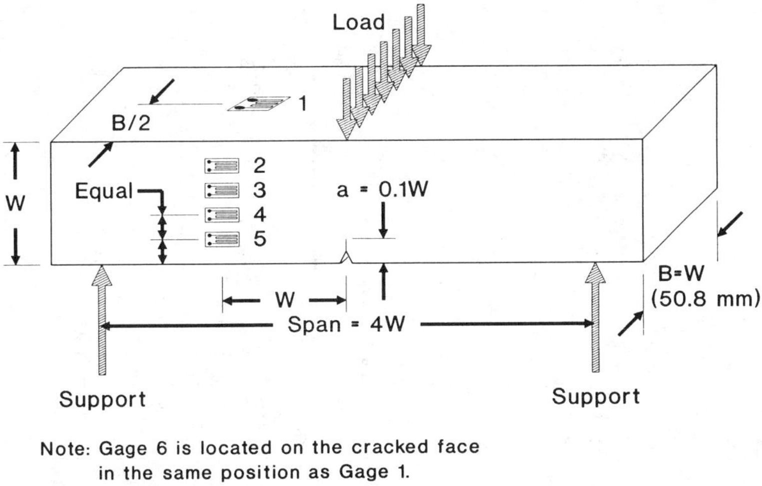

FIG. 1—*Diagram of shallow-cracked SE(B) specimen showing position of quarter-span strain gages used to estimate load.*

excursions characteristic of a steel on steel impact, these occurring as a result of loss of contact between the tup and the specimen. This technique has been employed previously by Madison and Irwin [*10*] and by Joyce and Hackett [*11*] using different absorber configurations.

Load Measurement—Load was measured by an array of six individual quarter bridge strain gages. These gages were positioned as illustrated in Fig. 1 to measure the bending strains along the quarter span position. Load was calculated from these bending strains using the following equation, which was derived using beam theory and was based on static equilibrium require-

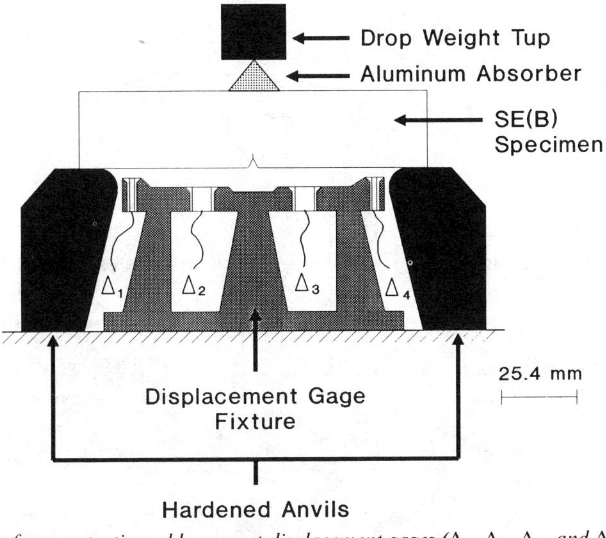

FIG. 2—*Diagram of noncontacting eddy current displacement gages (Δ_1, Δ_2, Δ_3, and Δ_4) positioned in the apparatus used to conduct SE(B) experiments.*

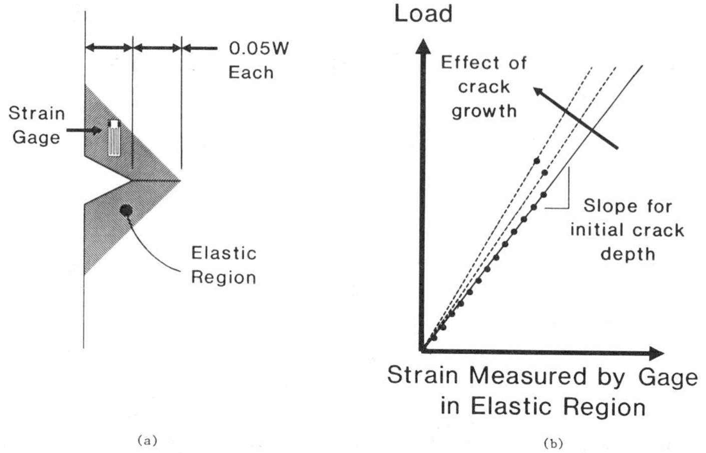

FIG. 3—(a) *Specimen region for strain measurement to infer crack initiation during shallow-cracked SE(B) experiments, and* (b) *anticipated load-strain response of a gage located in this region.*

ments, the boundary conditions, the position of the strain gage array relative to the loading points, and the assumption that plane sections remain plane.

$$P = \frac{-8 \cdot B}{S} \int_{W} \sigma(\epsilon) \cdot y dy \tag{2a}$$

where

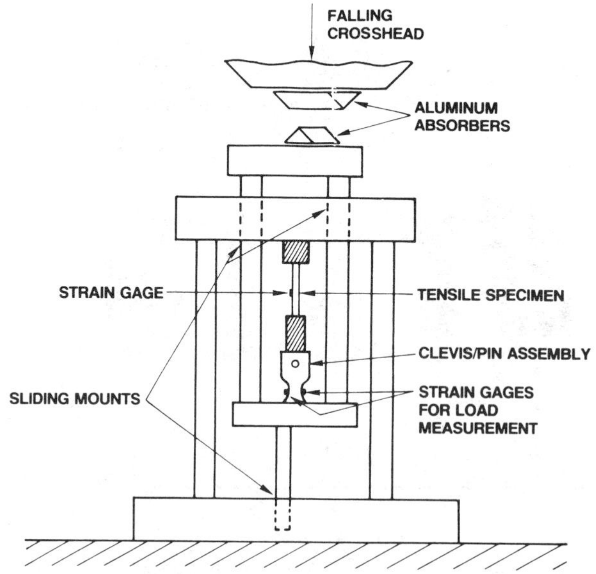

FIG. 4—*Apparatus used to conduct dynamic tension tests.*

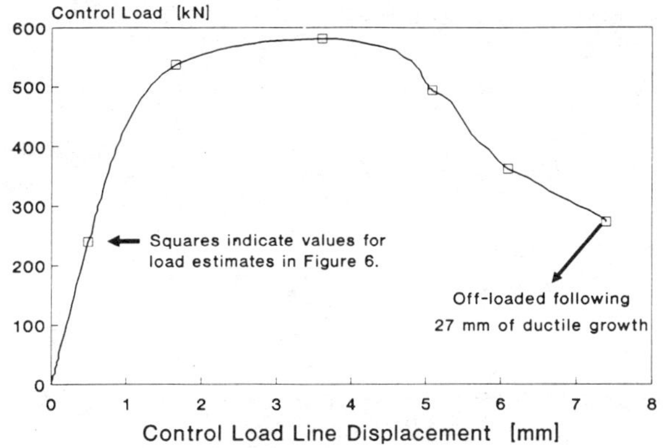

(a)

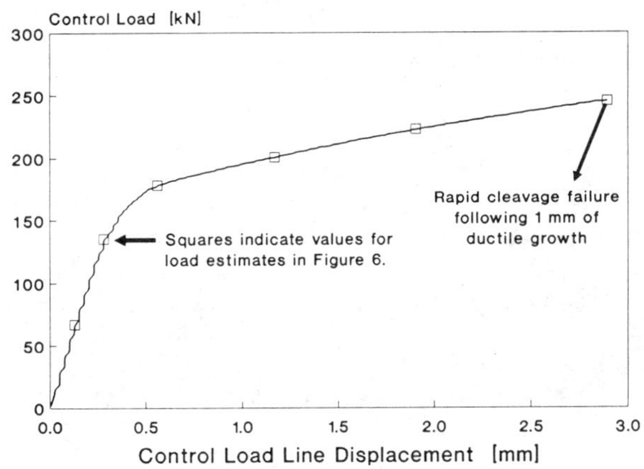

(b)

FIG. 5—*Control records of load versus load-line displacement recorded during quasi-static shallow-cracked SE(B) tests of* (a) *low and* (b) *high hardening steel alloy.*

$\sigma(\epsilon)$ = bending stress along quarter span based on measured strains,
W = specimen width,
S = span between supports = $4 \cdot W$, and
y = distance from neutral axis (positive toward tension side of the beam).

In Eq 2a, stresses were calculated from the measured strains using the following nonlinear stress-strain relationship:

$$\sigma(\epsilon) = \sigma_o \cdot 10^{(1/e + 0.3)} \tag{2b}$$

where
$e = \beta_o + \beta_1 \cdot \xi + \beta_2 \cdot \xi^2 + \beta_3 \cdot \xi^3 + \beta_4 \cdot \xi^4 + \beta_5 \cdot \xi^5 + \beta_6 \cdot \xi^6$,
$\xi = \epsilon_o/\epsilon$,
$\epsilon_o = \sigma_o/E$,

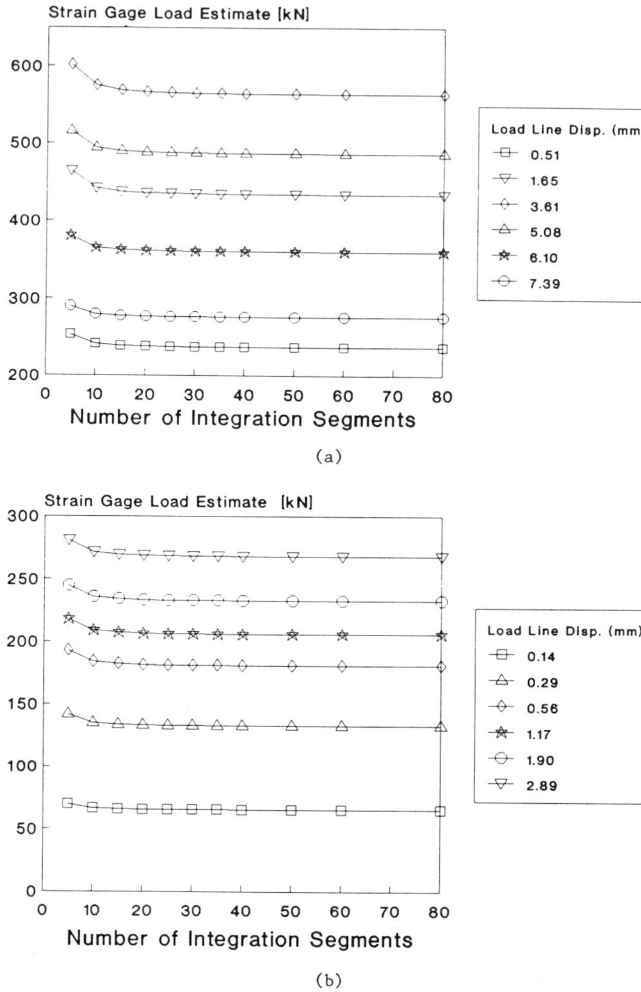

FIG. 6—*Effect of number of integration segments on load estimated from strain gage measurements using Eq 2a for* (a) *low and* (b) *high hardening steel.*

σ_o = 0.2% offset yield stress, and
β_i = empirical coefficients fit to experimental uniaxial engineering stress-strain data

This equation provides a close, closed form fit to experimental stress-strain data for the two steels investigated. It has no value other than as a fitting function and was used strictly for analytical convenience.

Load-Line Displacement Measurement—Load-line displacement measurements were made using four eddy current displacement gages. These noncontacting transducers were placed under the SE(B) specimen as shown in Fig. 2. The gages were offset from the load-line position to allow measurement of the bend angle imposed on the specimen using the following equation:

$$\theta = \mathrm{Tan}^{-1}\left(\frac{\Delta_2 - \Delta_1}{X}\right) + \mathrm{Tan}^{-1}\left(\frac{\Delta_3 - \Delta_4}{X}\right) \tag{3}$$

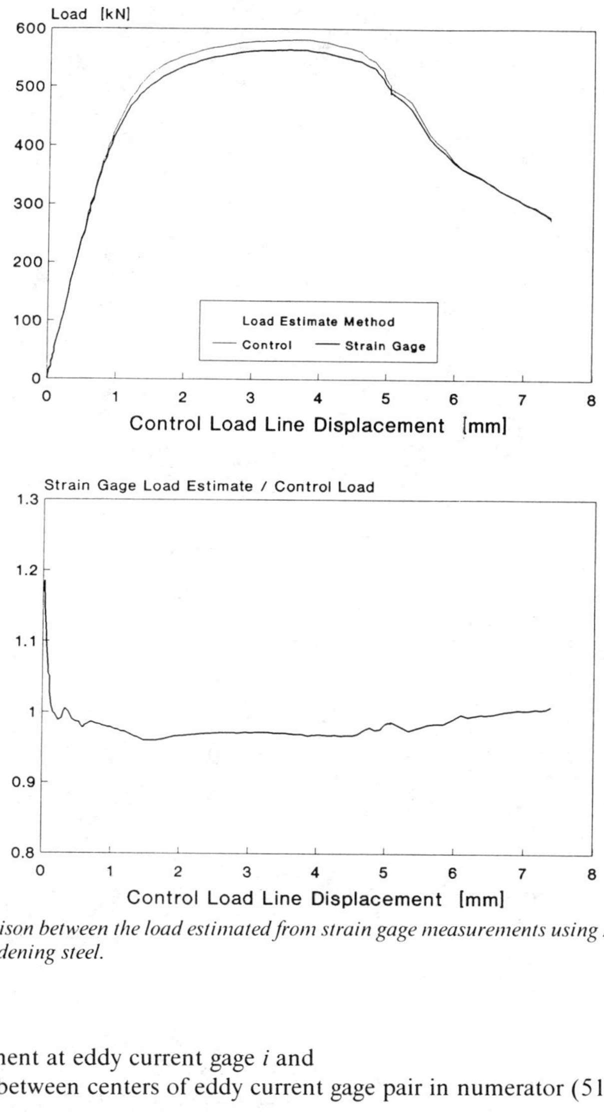

FIG. 7—*Comparison between the load estimated from strain gage measurements using Eq 2a and the control load for the low hardening steel.*

where

Δ_i = displacement at eddy current gage i and

X = distance between centers of eddy current gage pair in numerator (51.6 mm).

This value of θ was used to calculate load-line displacement as follows:

$$\Delta_{LL} = \frac{S}{2} \cdot \text{Tan}\ (\theta/2) \tag{4}$$

By basing the load-line displacement on differential displacement values ($\Delta_2 - \Delta_1$ and $\Delta_3 - \Delta_4$), displacement as a result of elastic compression of the loading anvils is eliminated because this component will be equal in all four individual displacement measurements. Equation 4 was derived by assuming rigid rotation of the specimen arms, a common assumption in fracture experimentation. Results presented in a following section demonstrate that this is incorrect for shallow flaws. However, the rigid rotation formula serves as a useful starting point and, lacking additional information, a more elaborate formula is not justified.

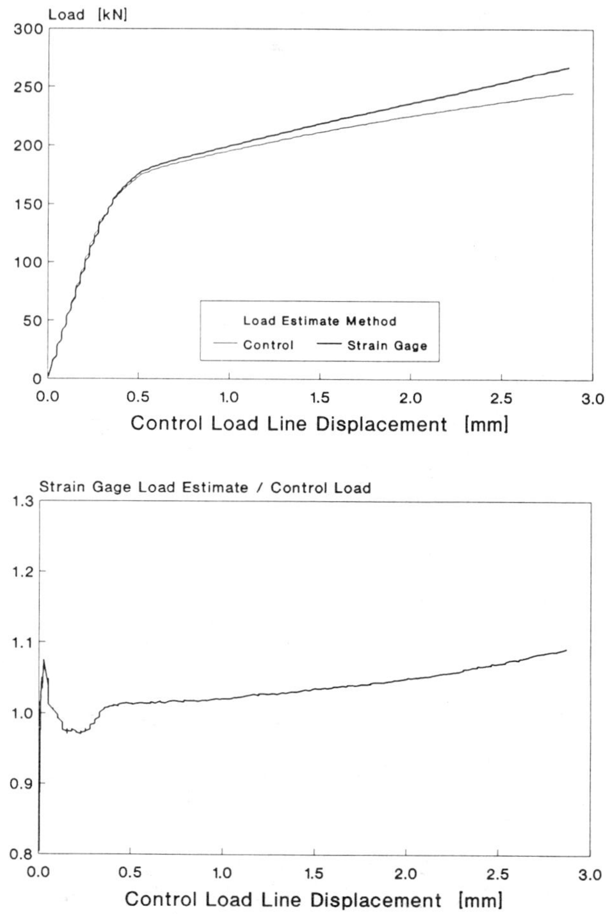

FIG. 8—*Comparison between the load estimated from strain gage measurements using Eq* 2a *and the control load for the high hardening steel.*

Crack Initiation Detection—During an impact experiment, conventional methods used in quasi-static experimentation to determine the time at which crack blunting stops and crack growth begins are generally not applicable. Unloading compliance is clearly not an option, and the direct current potential drop (DCPD) technique, although useful for nonferrous alloys, cannot be used for ferrous materials because of the magneto-elastic spike demonstrated by Hackett et al. [*12*] to swamp the crack growth induced potential change. The alternating current potential drop (ACPD) technique shows promise [*13*] but suffers from analytical complexity. In this investigation, placing small strain gages in close proximity to the crack tip was investigated as a simple alternative that would be generically applicable to both ferrous and nonferrous materials.

In developing a candidate procedure, it was desired to locate a position at which strain could be measured such that there would be an inflection point in a plot of load versus strain indicative of crack initiation. To this end, the elastically loaded region behind the fatigue crack tip, illustrated schematically in Fig. 3*a*, was focused on. In this region, strains increase in proportion to the applied load, the proportionality constant being determined by the material elastic modulus,

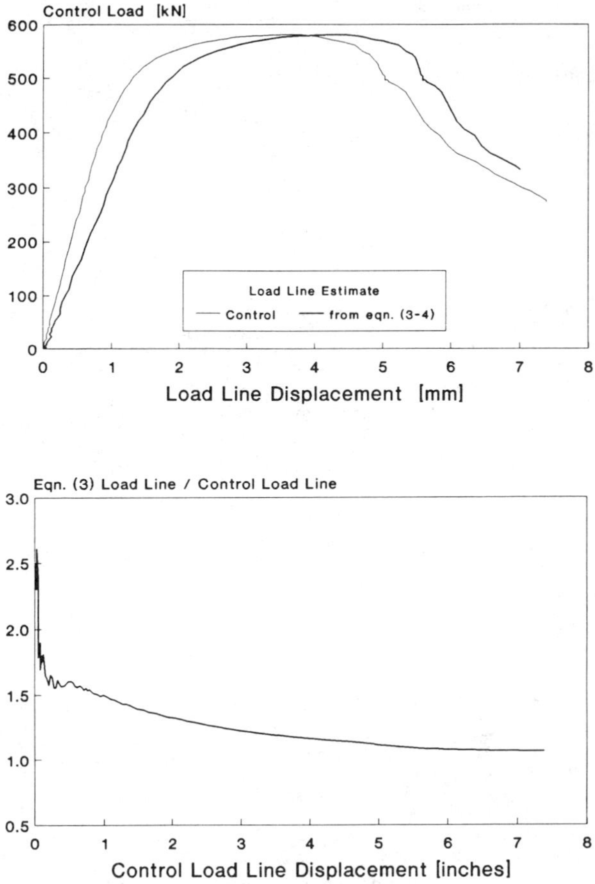

FIG. 9—*Comparison between the load-line displacement estimated from the eddy current displacement measurements using Eqs 3 and 4 and the control load-line displacement for the low hardening steel.*

the orientation of the gage, and the specimen geometry. When the crack initiates, the specimen geometry changes and the proportionality constant increases, more load being needed to cause the same amount of strain at the gage location. This concept is illustrated schematically in Fig. 3*b*. In practice, the strain reversal shown in Fig. 3*b* is probably not as sharp because of the gradual nature of a plastic crack-tip blunting process; however, a reversal in the load versus gage strain trace (or perhaps a deviation from linearity) should still be related to the onset of crack growth.

Tension Testing

Tension tests were conducted on the two materials at room temperature (+23°C) to determine the stress-strain behavior for the SE(B) specimen load analysis. Strains were measured with two high-elongation strain gages; these gages were shunt calibrated and conditioned through a high-frequency amplifier having a maximum frequency response of 50 kHz. Quasi-static tests

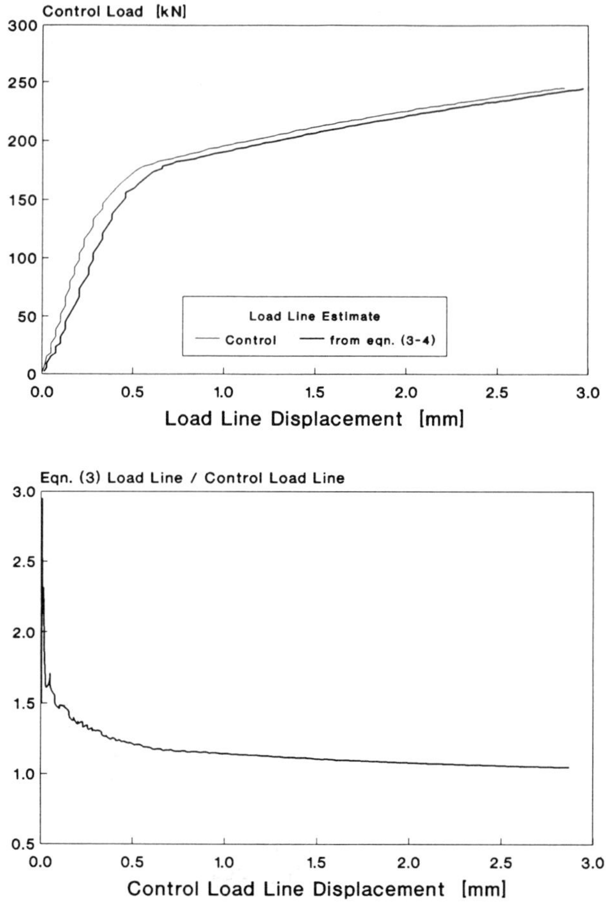

FIG. 10—*Comparison between the load-line displacement estimated from the eddy current displacement measurements using Eqs 3 and 4 and the control load-line displacement for the high hardening steel.*

were conducted in a screw-driven tensile machine at a constant crosshead rate; load was measured using a calibrated load cell. High-rate tests were conducted in a drop tower using a special procedure described below.

High-rate tension tests of the low-hardening steel were conducted using the dynamic loading rig illustrated in Fig. 4. This rig included a half bridge strain gage load cell that was calibrated versus the load cell in a screw-driven tensile machine. After load calibration, the tensile rig was placed in a drop tower where the load was applied by a free-falling 669-kg crosshead dropped from 0.61 m to achieve a displacement rate of 3.45 m/s. The difference between this displacement rate and that applied to the SE(B) specimens (4.88 m/s) was not considered significant because of the small effect of strain rate on the tensile properties of a similar alloy reported by Kirk and Hackett [*2*]. Two crossed triangular wedges of soft aluminum were placed between the falling crosshead and the top of the tensile rig to allow a smooth load transfer between the falling crosshead and the rig. Load and strain signals were recorded on a four-channel digital oscilloscope triggered by means of a fiber-optic light beam broken by a flag attached to the falling crosshead.

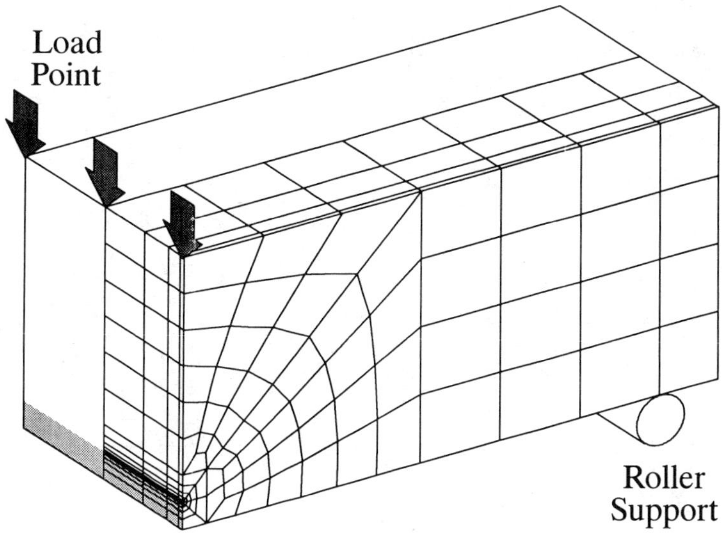

FIG. 11—*Finite element model of the shallow-cracked SE(B) specimen.*

Results and Discussion

Quasi-static Experiments

One shallow-cracked SE(B) specimen of each material was tested at a quasi-static loading rate (ram speed < 0.01 mm/s or 4×10^{-4} in./s) so that control measurements of load and load-line displacement could be made. Control load measurements were made using a calibrated load cell, whereas control load-line displacement measurements were made using a flex bar [*14*]. Comparisons between the load and load-line displacement estimated using the candidate procedures and these control values are made in the following sections.

Load—Strains were measured at six equally spaced locations along the quarter span (see Fig. 1) so that load could be estimated by integrating the stress profile across the specimen width using Eq 2. Numerical integration of this equation was achieved by partitioning the quarter span into a finite number of segments. Although the number of segments used is arbitrary, insufficient discretization will bias the load estimate. To determine the number of segments needed to minimize this error, the load at six points during each test, as indicated on Fig. 5, was estimated using various numbers of integration segments. The results of these calculations (Fig. 6) indicate that once the width of the specimen is partitioned into 40 or greater equal segments, the load estimated using Eq 2 is no longer influenced significantly. For this reason, 40 integration segments were used in all subsequent load estimates made by numerically integrating Eq 2.

Figures 7 and 8 compare the control and strain gage load estimates for the two specimens tested. These data show that once a slight load is placed on the specimens (50 to 100 kN), the strain gage load estimate stays between 96 and 101% of the control load for the lower hardening material and between 97 and 110% of the control load for the higher hardening material. While it would be preferable to have better agreement between the estimated and control loads for the higher hardening material, it will be demonstrated in a later section that this load error magnitude does not significantly affect the accuracy with which applied J can be calculated from these data. This indicates that this procedure for estimating the applied load from strain gage data has acceptable accuracy for load measurement during impact elastic-plastic fracture experiments.

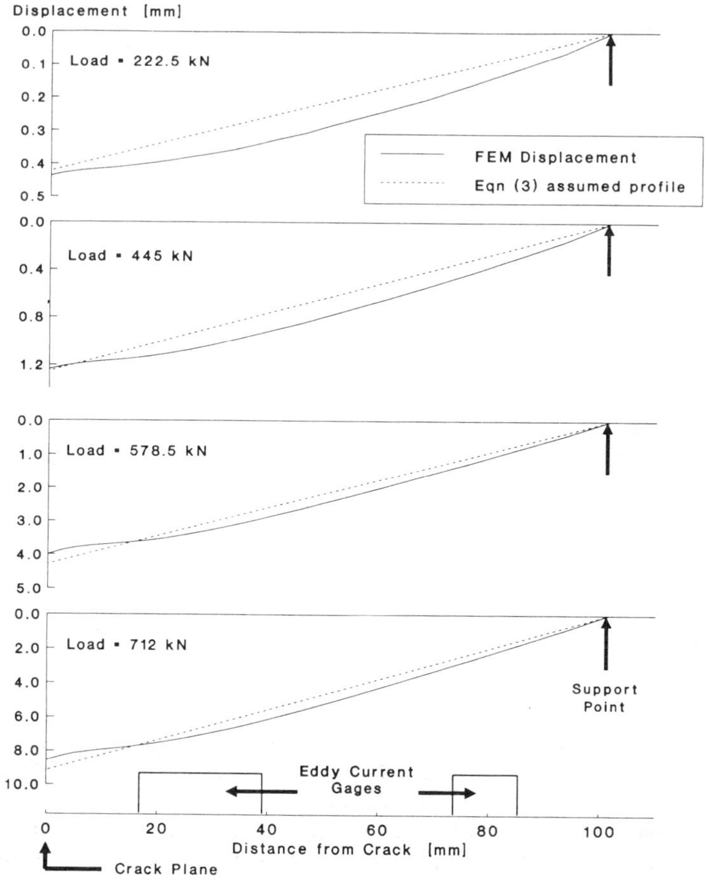

FIG. 12—*Comparison of the displaced profile at the mid-thickness of the cracked side of the SE(B) specimen predicted using finite elements and that assumed in the derivation of Eq 3 at various load levels.*

Load-Line Displacement—In Figs. 9 and 10, the control load-line displacement measured using the flex bar is compared to the load-line displacement estimated by Eqs 3 and 4 using the four-eddy current gages for the two specimens tested. These data indicate that the eddy current gage estimate of load-line displacement considerably exceeds the control measurement, the error being most significant at small displacements. To investigate the cause of this discrepancy, a three-dimensional finite-element model of the shallow-cracked SE(B) specimen was constructed using 324 20-node isoparametric elements (Fig. 11). Symmetry of the loading and constraints permitted modeling of only one quarter of the specimen. Elements incident on the crack front were degenerated into triangular prisms with the coincident nodes free to displace independently. This modeling procedure provides a strain singularity inversely proportional to distance from the crack tip ($1/r$) at all points along the crack front corresponding to element corner nodes. Reduced integration was used for all elements. Plasticity was modeled using incremental theory with a von Mises yield surface, an associated flow rule, and isotropic hardening. The material was assumed to have a bilinear uniaxial stress-strain curve having a slope of 206 850 MPa before yield, a slope of 6895 MPa after yield, and a yield stress of 675.7 MPa. Conven-

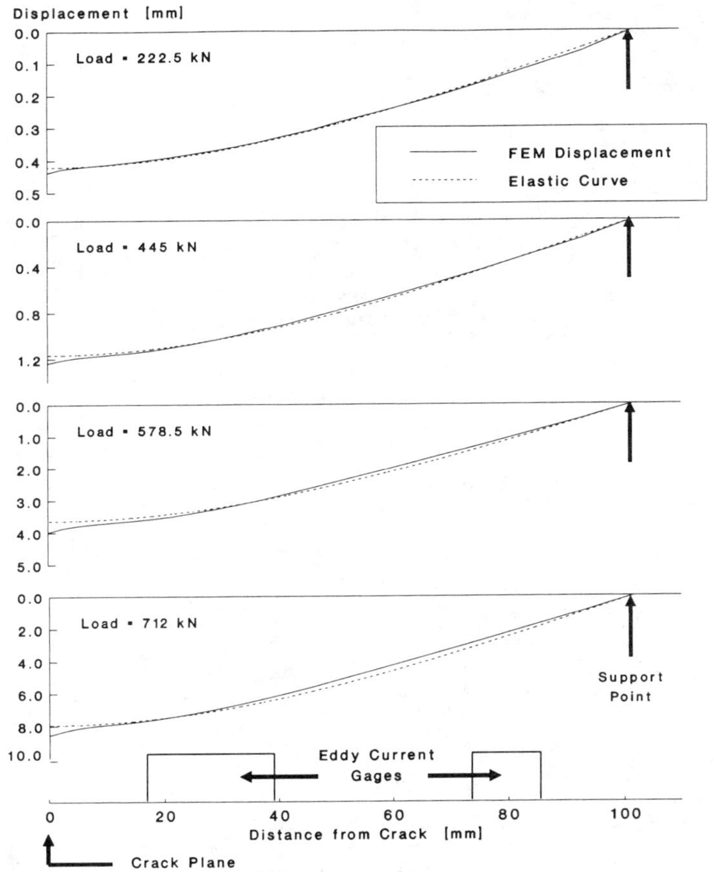

FIG. 13—*Comparison of the displaced shape at the mid-thickness of the cracked side of the SE(B) specimen predicted using finite elements and that of a beam elastically loaded in three-point bending.*

tional, linear strain-displacement relations were employed by the computations. The finite-element models were loaded by enforcing uniform vertical displacement of the nodes along a line at the top of the uncracked ligament and applying a load to the center point of this line. These computations were performed using POLO-FINITE software.

Figure 12 shows the deformed shape at the mid-thickness of the specimen's cracked side at various load levels as determined from this finite-element model. These data indicate that the cracked side of the bar deforms into a curve, rather than as a straight line. This curved deformation causes Eqs 3 and 4 to overestimate load-line displacement because the curved profile of the specimen is more displaced from the zero position than the straight line deformation assumed in deriving Eq 3. To improve the accuracy of the load-line displacement estimated by Eqs 3 and 4, Eq 3 could be corrected for the error caused by the curved deformation, provided the magnitude of the error at each gage is easily calculable based on experimental measurements. Figure 13 shows that the deformed shape of the specimen predicted by the finite-element analysis can be reasonably approximated by the cubic curvature of a beam elastically loaded in three-point bending even after significant yielding. This suggests that better agreement with the control values of load-line displacement could be achieved by modifying Eq 3 as follows:

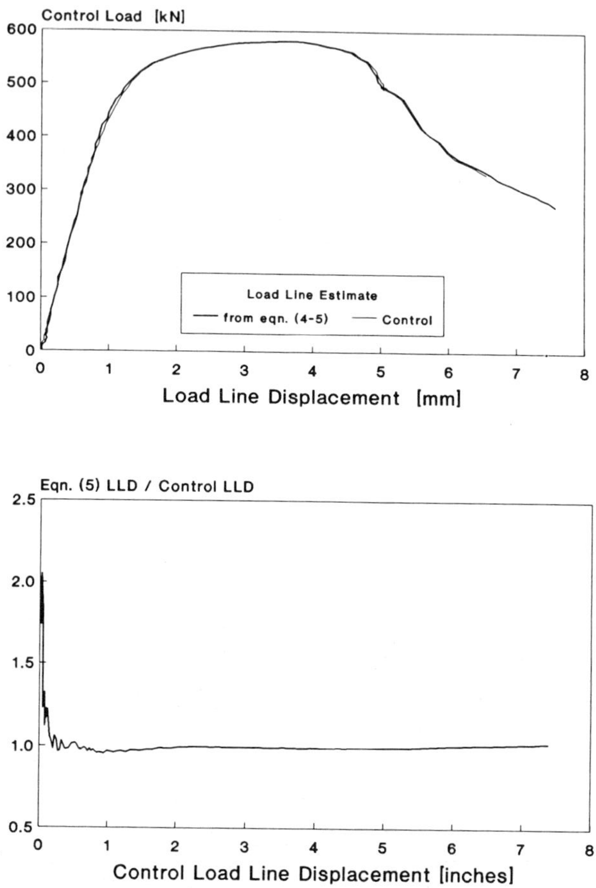

FIG. 14—*Comparison between the load-line displacement estimated from the eddy current displacement measurements using Eqs 4 and 5 and the control load-line displacement for the low hardening steel.*

$$\theta^* = \text{Tan}^{-1}\left(\frac{\Delta_2 - \alpha\Delta_2^e - (\Delta_1 - \alpha\Delta_1^e)}{X}\right) + \text{Tan}^{-1}\left(\frac{\Delta_3 - \alpha\Delta_3^e - (\Delta_4 - \alpha\Delta_4^e)}{X}\right) \tag{5}$$

where

α = empirical factor,

Δ_i^e = elastic displacement at eddy current gage i calculated using the following equation:

$$\Delta_i^e = \frac{P}{48 \cdot EI}(3S^2x - 4x^3)$$

In this equation, x is the distance from the nearest roller support to the center of the eddy current gage. Thus, for Gages 1 and 4, x = 22.1 mm, whereas for Gages 2 and 3, x = 73.66 mm.

and using θ^* rather than θ in Eq 4 to estimate load-line displacement. Figures 14 and 15 show that load-line displacement estimated using θ^* is in considerably better agreement with the

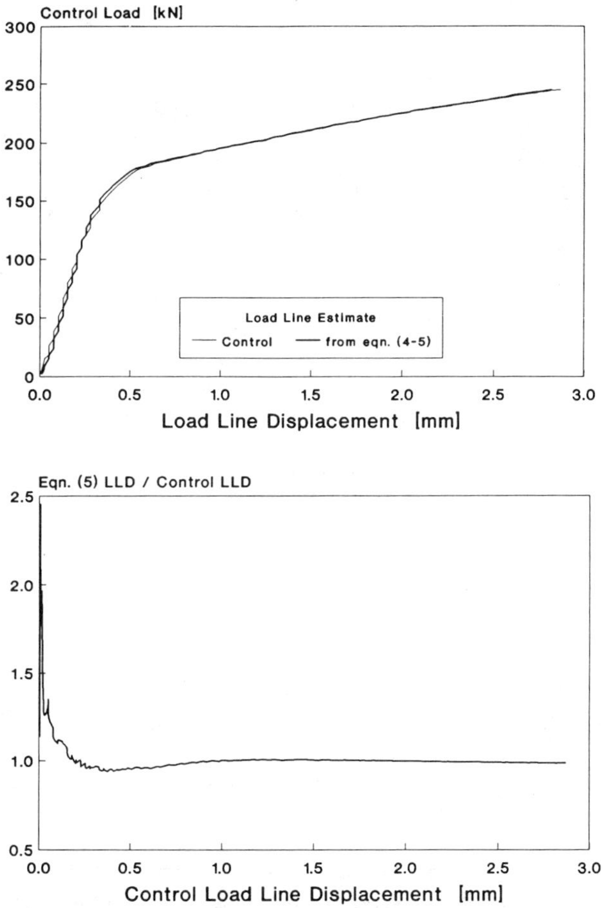

FIG. 15—*Comparison between the load-line displacement estimated from the eddy current displacement measurements using Eqs 4 and 5 and the control load-line displacement for the high hardening steel.*

control measurements (within ± 6% for both steels) than load-line displacement estimated using θ. In these figures, the α value was selected to force the initial load versus load-line displacement slope match that expected for the initial precrack depth; this slope being calculated using the elastic compliance equations presented in Ref *15*. This makes the choice of α nonarbitrary and further allows it to be made without the need for a control load-line displacement to match, which is the case at impact loading rates. These data indicate that estimates of load-line displacement made using this procedure have acceptable accuracy for load measurement during impact elastic-plastic fracture experiments.

J *Calculation*—In the preceding sections, it was demonstrated that both load and load-line displacement can be determined using the candidate procedures to acceptable levels of accuracy for two materials of considerably different strain-hardening capacity. Because the result of a fracture-toughness test is an estimate of the fracture initiation toughness, it is of interest to determine how load and load-line displacement estimation errors affect the accuracy with which J can be calculated using Eq 1. Figures 16 and 17 compare the J values calculated using the

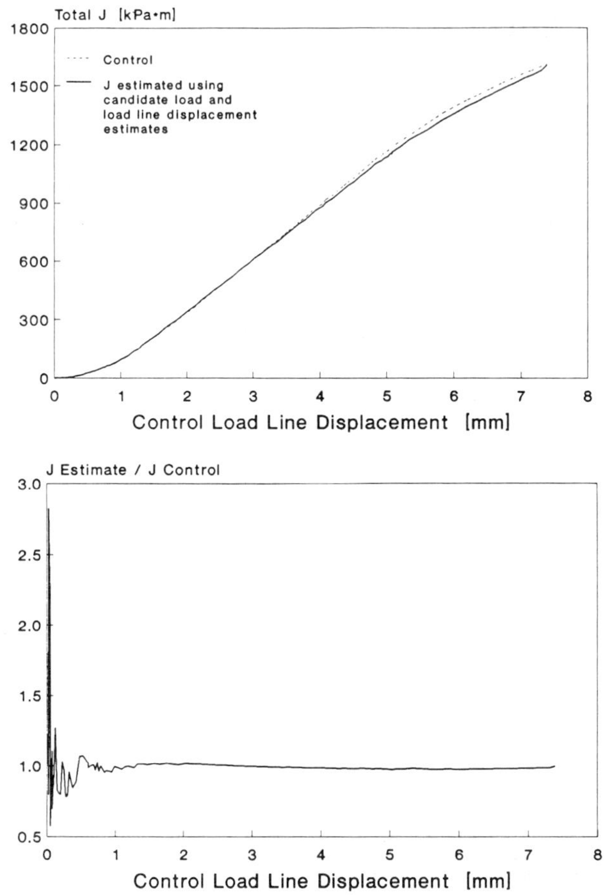

FIG. 16—*Comparison of applied* J *calculated from the reference load and load-line displacement estimates with that calculated from load and load-line displacement estimated using the candidate procedures for the low hardening steel.*

control load and load-line displacement data to *J* values calculated from load and load-line displacement estimated using the candidate procedures discussed above. Poor *J* estimates occurred at small displacements/loads because the transducers were calibrated over a much larger range required to capture the full elastic-plastic response of these specimens. However, at larger displacements where crack initiation was expected, *J* values were estimated to within ± 3% of the control values using the candidate load and load-line displacement estimation procedures. This indicates that these procedures have acceptable accuracy for estimation of applied *J* values during impact elastic-plastic fracture experiments.

Impact Testing

A shallow-cracked SE(B) specimen of the lower hardening steel was tested in a drop tower at a terminal impact velocity of 4.88 m/s to a total displacement of 4.3 mm at which point the falling head was arrested on rigid stops, allowing the specimen to unload. Post-test examination of the

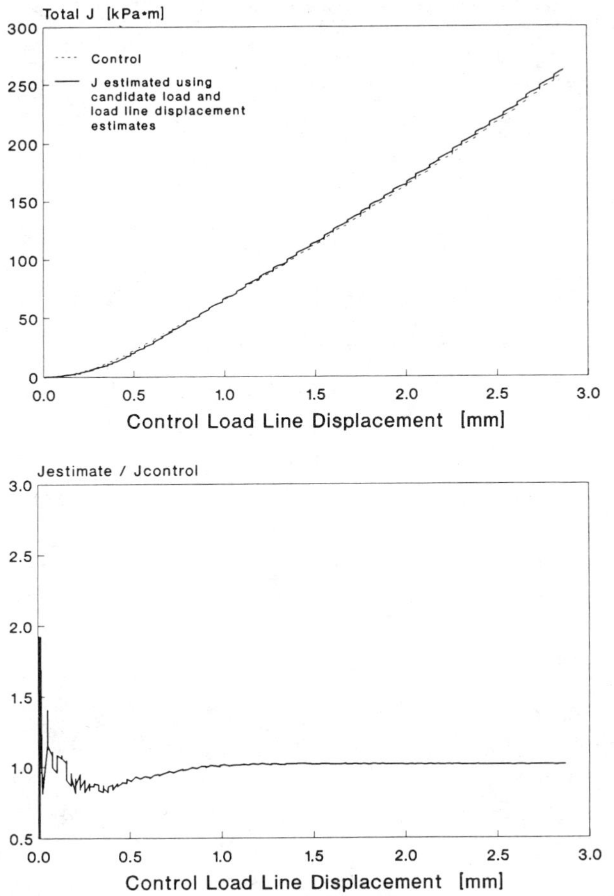

FIG. 17—*Comparison of applied* J *calculated from the reference load and load-line displacement estimates with that calculated from load and load-line displacement estimated using the candidate procedures for the high hardening steel.*

fracture surface indicated that this amount of deformation resulted in 2.1 mm of ductile crack growth at the specimen midplane. Figure 18 shows the variation of load and load-line displacement with time for this experiment, these values having been estimated from the strain and displacement gage data as discussed previously. Figure 19 shows the load versus load-line displacement curve, a value of $\alpha = 0.60$ having been used in Eq 5 to bring the initial load versus load-line displacement slope into agreement with that expected based on the initial precrack length.

The data presented in Fig. 18 demonstrate the adequacy of these strain and displacement measurements to resolve load and load-line displacement variations at microsecond intervals, such a response being more than adequate for this type of impact experiment in which the maximum load plateau is reached 3 ms after initial load application. However, because Eq 1 considers only the effects of deformation energy (W) and not of kinetic energy (K) in the calculation of J, the error introduced by this simplifying assumption needs to be determined before these load and load-line displacement values can be used with confidence to calculate an

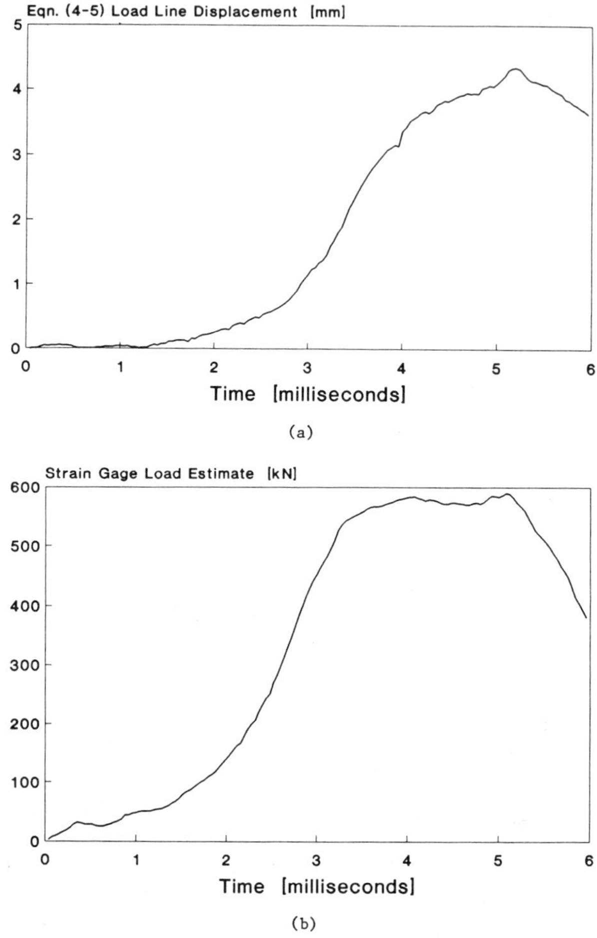

FIG. 18—(a) *Load-line displacement versus time and* (b) *load versus time response of a shallow crack SE(B) specimen of the low hardening steel tested in a drop tower at 4.88 m/s.*

initiation fracture-toughness value. Nakamura et al. [*16*] developed a model from which the ratio of kinetic energy to deformation energy (K/W) can be calculated as a function of time based on load-line displacement measurements. This model is as follows:

$$\frac{K}{W} = \left[\left(\frac{S \cdot B \cdot E \cdot C_s}{2 \cdot W}\right)^{1/2} \cdot \frac{W}{c_0} \cdot \frac{\dot{\Delta}_{LL}(t)}{\Delta_{LL}(t)}\right]^2 \tag{6}$$

where

C_s = elastic compliance and

c_0 = longitudinal bar wave speed.

Nakamura et al. demonstrated that Eq 6 provides good estimates of the K/W ratio for an SE(B) specimen with an a/W of 0.5 by comparing the predictions of Eq 6 to K/W values determined directly from a dynamic finite-element model of the SE(B) specimen. This is considered ade-

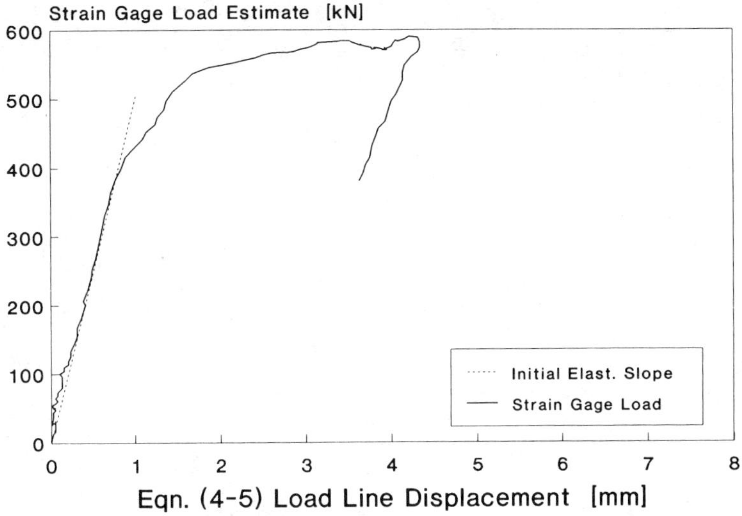

FIG. 19—*Load versus load-line displacement response of a shallow crack SE(B) specimen of the low hardening steel tested in a drop tower at 4.88 m/s.*

quate justification for use of Eq 6 to analyze data from a shallow-notched SE(B) ($a/W \approx 0.1$) because the assumptions made in the derivation of Eq 6 (the presence of the crack was ignored and it was assumed that plastic deformation is insignificant for all times when K/W exceeds unity) are, if anything, better satisfied by the shallow crack case. Figure 20 shows the variation of K/W with time for the impact-loaded SE(B) specimen; these data indicate that the K/W ratio is

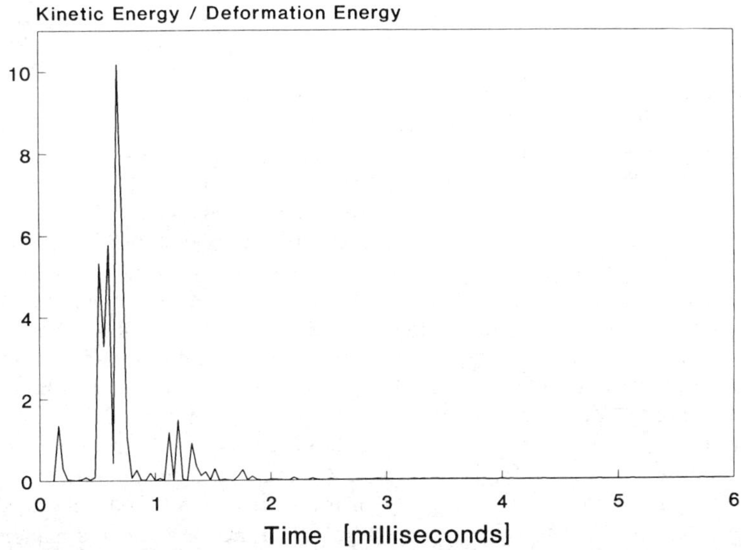

FIG. 20—*Variation of the ratio of kinetic energy to deformation energy with time in a shallow crack SE(B) specimen of the low hardening steel tested in a drop tower at 4.88 m/s.*

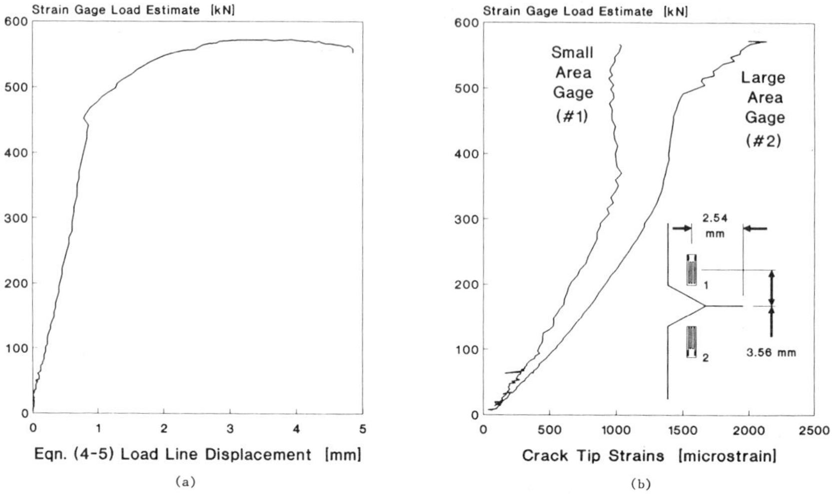

FIG. 21—*Test record from an impact SE(B) test of the low hardening steel that resulted in 4.70 mm of crack growth at the specimen midplane.* (a) *Load versus load-line displacement data.* (b) *Load versus crack-tip region strain.*

quite small (less than 0.1) for all times after 1.75 ms. At this time, the specimen is still subjected to globally elastic loading (load < 100 kN) and is nowhere near crack initiation. This demonstrates that ignoring the effects of kinetic energy in calculating the crack initiation toughness of a shallow-cracked SE(B) specimen loaded at this rate will not significantly influence the J_i estimate because the amount of kinetic energy in the specimen is small compared to the amount of deformation energy from very early on in the loading history.

Crack Initiation Detection

To test the candidate procedure described in the experimental procedure section, a shallow-cracked SE(B) specimen of the lower hardening steel was instrumented with two foil strain gages bonded close to the crack tip in addition to the load and load-line displacement instrumentation described earlier. Gages of two different grid sizes were attached at the same nominal position and orientation (2.54 mm behind the crack tip, 3.56 mm above the crack line, perpendicular to the crack line) with respect to the crack tip. The specimen was allowed to displace 4.85 mm at the load line before the falling crosshead was arrested onto rigid stops, a displacement that resulted in 4.70 mm of crack growth at the specimen mid-plane. Figure 21 shows the resulting load versus load-line displacement and load versus crack-tip strain traces. Only the strain measured by smaller gage (gage length of 0.79 mm, grid area of 0.64 mm^2) showed a reversal with load, whereas the larger gage (gage length of 1.57 mm, grid area of 2.46 mm^2) measured systematically larger strains that increased monotonically with load. It was reasoned that both gages had been placed too close to the elastic-plastic boundary, and that a better defined strain reversal could be achieved by moving the gages further into the elastic region. The new gage position was 3.56 mm behind the crack tip and 2.54 mm above the crack line; this repositioning precluded further use of the large strain gage. A specimen so instrumented was allowed to displace 3.33 mm at the load line, this resulting in 1.52 mm of crack growth at the specimen mid-plane. Figure 22 shows the resulting load versus load-line displacement and load versus crack-tip strain traces; clear reversals of strain with load being observed in this experiment. At this point, it was envi-

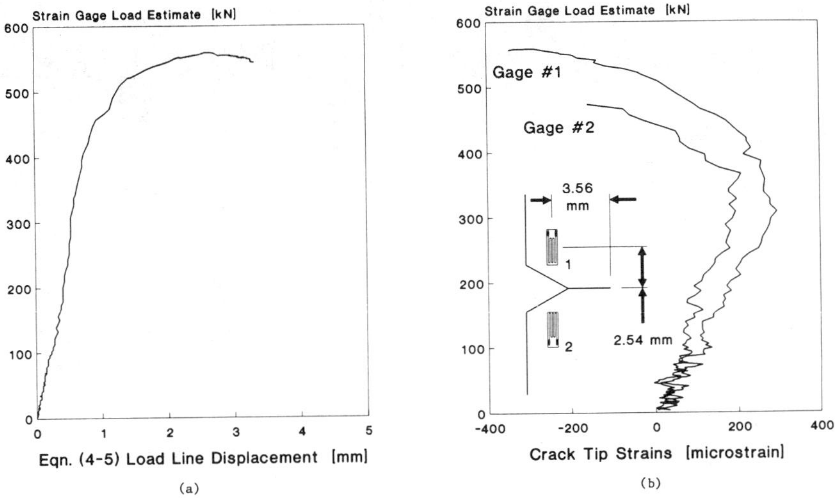

FIG. 22—*Test record from an impact SE(B) test of the low hardening steel that resulted in 1.52 mm of crack growth at the specimen midplane.* (a) *Load versus load-line displacement data.* (b) *Load versus crack-tip region strain.*

sioned that several more impact experiments would be conducted at progressively lower levels of total load-line displacement until no crack growth was observed. In practice, the arrestor mechanism on the impact tester could not be controlled to the degree needed to achieve this goal; after two additional impact experiments that showed 1.65 to 2.03 mm of crack growth at the specimen mid-plane and clear strain reversals, this approach was abandoned.

As an alternative to impact experimentation, two quasi-static experiments were conducted to allow close control on the displacement at which the test was stopped. One test was halted after the load versus strain reversal, whereas the other was halted just as the reversal was about to occur; the data from these experiments are presented in Figs. 23 and 24. After heat tinting and breaking open at liquid nitrogen temperatures, neither specimen showed macroscopic evidence of crack initiation. However, a scanning electron microscope examination of the fracture surfaces revealed pockets of ductile growth between the fatigue crack and the final cleavage fracture on both specimens (Fig. 25). These data suggest that the reversal in the load versus crack-tip strain plots of Figs. 22 through 24 is closely associated with crack initiation from the fatigue crack, making this strain measurement useful for inferring the time at which crack initiation occurs. This information, combined with the load and load-line displacement data, allows an initiation fracture-toughness value to be estimated from the results of impact tests of shallow-cracked SE(B) specimens. Note, however, that the applicability of this method to different materials is likely to depend on both the strain-hardening exponent and the initiation fracture toughness of the material being tested; further work is required to quantify these effects.

Summary and Conclusions

In this investigation, techniques useful for estimating load, load-line displacement, and the time of crack initiation during impact tests of single-edge notched bend [SE(B)] specimens having shallow fatigue cracks ($a/W \approx 0.1$) were developed and validated. In particular, the following conclusions may be drawn:

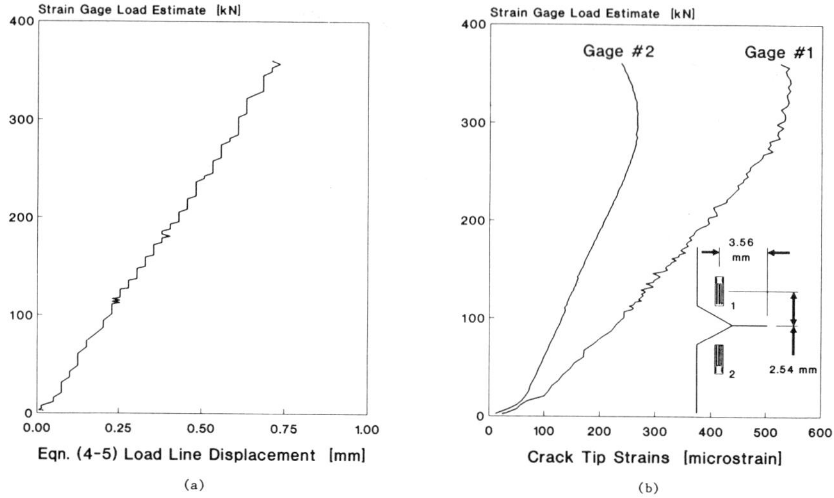

FIG. 23—*Test record from a quasi-static SE(B) test of the low hardening steel that did not result in macroscopically observable crack growth.* (a) *Load versus load-line displacement data.* (b) *Load versus crack-tip region strain.*

1. Estimates of the applied load on a shallow-cracked SE(B) specimen can be made by integrating the distribution of stress across the arm of the specimen midway between the loading and support points. Information regarding the variation of bending strain with position across the specimen arm and the uniaxial engineering stress-strain curve are needed to make this estimate. These estimates of load compared favorably with control

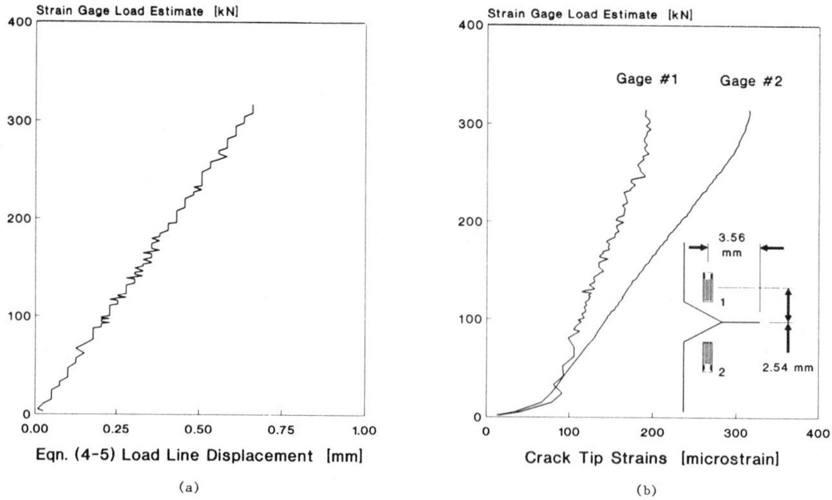

FIG. 24—*Test record from a quasi-static SE(B) test of the low hardening steel that did not result in macroscopically observable crack growth.* (a) *Load versus load-line displacement data.* (b) *Load versus crack-tip region strain.*

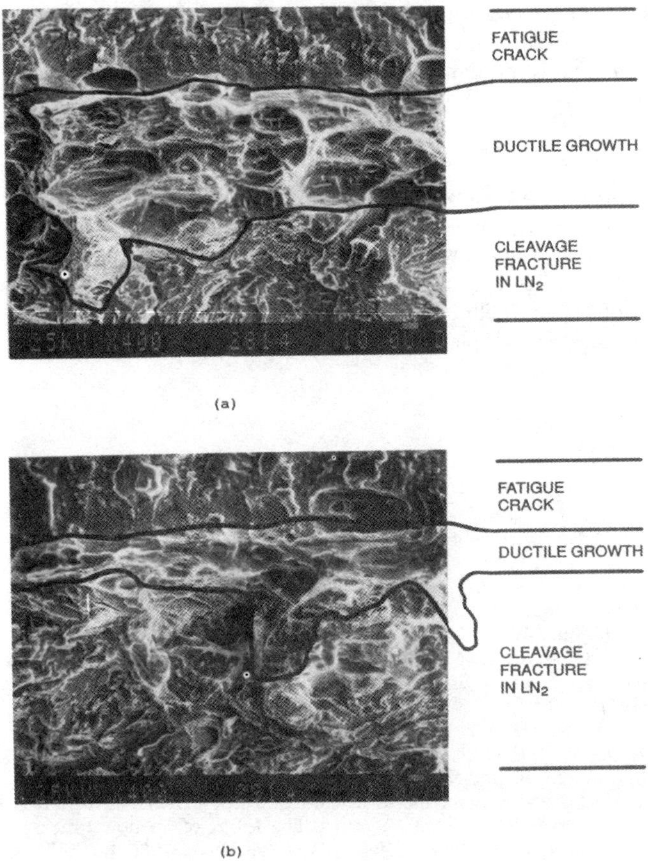

FIG. 25—*Microscopic ductile crack growth that resulted from the loading shown in* (a) *Fig. 23 and in* (b) *Fig. 24.*

measurements made during quasi-static SE(B) tests of two steel alloys, one having a low and one having a high strain-hardening exponent.

2. The load-line displacement experienced by a shallow-cracked SE(B) specimen can be estimated from displacements measured at four locations on the cracked face of the specimen, two on each side of the crack. These measurements were used to calculate the bend angle experienced by each half of the specimen, the load-line displacement then being estimated by assuming rigid rotation. Once corrected for the nonlinear deformation of the specimen using a simple formula derived from Euler beam theory, these estimates compared favorably with control measurements made during quasi-static SE(B) tests of two steel alloys, one having a low and one having a high strain-hardening exponent.
3. It is appropriate to ignore the effects of kinetic energy when calculating an initiation fracture-toughness value from load and load-line displacement data for shallow-cracked SE(B) specimens loaded at up to 4.88 m/s because this energy is a small fraction of the deformation energy in the specimen at the time of crack initiation. This finding was demonstrated by dynamic testing of a low hardening steel alloy.
4. The occurrence of a reversal in a plot of applied load versus strain normal to the crack

plane measured in the elastically loaded region behind the crack tip in a shallow-cracked SE(B) specimen indicates crack initiation from the fatigue crack. This finding was demonstrated by quasi-static and dynamic testing of a low hardening steel alloy.

Acknowledgments

The authors are pleased to acknowledge the able technical assistance rendered by M. Keppel, S. Mikalac, and J. Hein. Additionally, the many helpful discussions with M. G. Vassilaros and E. M. Hackett are gratefully acknowledged. We are especially grateful to E. M. Hackett for his kind provision of the tensile data for the low hardening material. This work was performed as part of the Office of Naval Technology Materials Block, which is administered at the David Taylor Research Center by I. Caplan.

References

[*1*] Sumpter, J. D. G., "Prediction of Critical Crack Size in Plastically Strained Welded Panels," in *Nonlinear Fracture Mechanics: Volume II—Elastic-Plastic Fracture, ASTM STM 995*, J. D. Landes, A. Saxena, and J. G. Merkle, Eds., American Society for Testing and Materials, Philadelphia, 1989, pp. 415–432.
[*2*] Kirk, M. T. and Hackett, E. M., "Fracture Behavior Prediction for Rapidly Loaded Surface Cracked Specimens," in *Surface Crack Growth: Models, Experiments, and Structures, ASTM STP 1060*, W. G. Reuter, J. H. Underwood, and J. C. Newman, Jr., Eds., American Society for Testing and Materials, Philadelphia, 1990, pp. 142–151.
[*3*] Matsoukas, G., Cotterell, B., and Mai, Y. W., "Hydrostatic Stress and Crack Opening Displacement in Three-Point Bend Specimens with Shallow Cracks," *Journal of Mechanics and Physics of Solids*, Vol. 34, No. 5, 1986, pp. 499–510.
[*4*] Zhang, D. Z. and Wang, H., "On the Effect of the Ratio a/W on the Values of δ_i and J_i in a Structural Steel," *Engineering Fracture Mechanics*, Vol. 26, No. 2, 1987, pp. 247–250.
[*5*] Matsoukas, G., Cotterell, B., and Mai, Y. W., "Crack Opening Displacement and Hydrostatic Stress," Department of Mechanical Engineering, University of Sydney, Sydney, Australia.
[*6*] Matsoukas, G., Cotterell, B., and Mai, Y. W., "The Effect of Shallow Cracks on Crack Opening Displacement," *Engineering Fracture Mechanics*, Vol. 24, No. 6, 1986, pp. 837–842.
[*7*] You, C. P. and Knott, J. F., "Effects of Crack Shape on Fracture Toughness in a High Strength Structural Steel," *Engineering Fracture Mechanics*, Vol. 24, No. 2, 1986, pp. 291–305.
[*8*] Sumpter, J. D. G., "J_c Determination for Shallow Notch Welded Bend Specimens," *Fatigue Fracture Engineering Material Structures*, Vol. 10, No. 6, 1987, pp. 479–493.
[*9*] Sorem, W. A. et al, "An Analytical Comparison of Short Crack and Deep Crack CTOD Fracture Specimens of an A36 Steel, presented at the ASTM 21st National Symposium on Fracture Mechanics held in Annapolis, MD, 28–30 June 1988.
[*10*] Madison, R. B. and Irwin, G. R., "Dynamic K_c Testing of Structural Steel," *Journal of the Structural Division, Proceedings of the American Society of Civil Engineers*, Vol. 100, No. ST7, July 1974, pp. 1331–1349.
[*11*] Joyce, J. A. and Hackett, E. M., "An Advanced Procedure for *J-R* Curve Testing Using a Drop Tower," in *Nonlinear Fracture Mechanics: Volume I—Time Dependent Fracture, ASTM STP 995*, A. Saxena, J. D. Landes, and J. L. Bassani, Eds., American Society for Testing and Materials, Philadelphia, 1989, pp. 298–317.
[*12*] Hackett, E. M., Kirk, M. T., and Hays, R. A., "An Evaluation of *J-R* Curve Testing of Nuclear Piping Materials Using the Direct Current Potential Drop Technique," NUREG/CR-4540, U.S. Nuclear Regulatory Commission, Washington, DC, Aug. 1986.
[*13*] Joyce, J. A., "Application of Alternating Current Potential Difference to Drop Tower *J-R* Curve Measurement," NUREG/CR-4699, Aug. 1986.
[*14*] Kirk, M. T. and Hackett, E. M., "An Evaluation of *J-R* Curve Testing Using Three-Point Bend Specimens," in *Fracture Mechanics: Eighteenth Symposium, ASTM STP 945*, D. T. Read and R. P. Read, Eds., American Society for Testing and Materials, Philadelphia, 1988, pp. 347–373.
[*15*] Tada, H., Paris, P. C., and Irwin, G. R., *The Stress Analysis of Cracks Handbook*, Del Research Corp., 1973.
[*16*] Nakamura, T., Shih, C. F., and Freund, L. B., "Three-Dimensional Transient Analysis of a Dynami-

cally Loaded Three-Point-Bend Ductile Fracture Specimen," in *Nonlinear Fracture Mechanics: Volume I—Time Dependent Fracture, ASTM STP 995*, A. Saxena, J. D. Landes, and J. L. Bassani, Eds., American Society for Testing and Materials, Philadelphia, 1989, pp. 217–241.

DISCUSSION

John D. Landes[1] *(written discussion)*—In your plot of load versus displacement that had time in milliseconds included, why was the displacement versus time so irregular?

Mark T. Kirk (author's closure)—Impact load is applied in drop-weight testing through a pyramidal piece of soft aluminum. When the falling weight first touches the apex of the pyramid, there is virtually no cross-sectional area to resist the load. Therefore, the aluminum deforms, blunting the top of the pyramid and transmitting very little displacement to the specimen. This deformation continues as the crosshead falls, steadily increasing the load-bearing area at the top of the blunted pyramid. As this blunting increases the pyramid's axial stiffness, larger fractions of the applied displacement are transmitted through the pyramid to the specimen as the pyramid flattens. After approximately 2.7 ms, the aluminum is completely deformed and the specimen begins to move at the full velocity of the falling crosshead (4.88 m/s) as shown by the nearly constant variation of displacement with time between 2.7 and 4 ms. At approximately 4 ms, the falling crosshead strikes semi-rigid stops, which cause the deceleration depicted in Fig. 18*a* after this time.

Richard E. Johnson[2] *(written discussion)*—Professor Dodds, one of your coauthors, also was a coauthor of the Welding Research Council Bulletin 351 (Feb. 1990), along with William A. Sorem and Stanley T. Rolfe. In the papers that constitute WRC-351, it was noted in specimens similar to yours that when plasticity developed from relatively deep cracks (a/w = 0.2 and greater) it took the form of a plastic hinge extending to the opposite (uncracked) surface. For smaller cracks ($a/w < 0.2$), however, the plastic zone tended to extend from the crack tip *back* to the free surface (under tensile load) behind the crack. If you are aware of these observations, could you comment on their relevance to your work, please?

Mark T. Kirk (author's closure)—Multiple strain gages across the quarter span are needed to determine the applied load because of the gross section yield behavior of shallow-cracked SE(B) specimens to which Dr. Johnson alludes. In contrast to this behavior, plasticity is confined to the ligament of deeply cracked SE(B) specimens ($a/W > 0.3$). In this case, strains at the quarter span remain fully elastic with a linear distribution across the specimen width. Thus, Joyce and Hackett [*11*, main text] found that, for deeply cracked SE(B) specimens, the applied load could be determined using elastic beam theory formulas and the readings from two strain gages placed on the extreme outer fibers of the specimen at the quarter span. However, gross section yielding disturbs the linear variation of bending strain with position across the quarter span for shallow-cracked SE(B) specimens. This necessitates the use of multiple gages across the specimen width to define this variation accurately.

[1] University of Tennessee, 310 Perkins Hall, Knoxville, TN 37996-2030.
[2] U.S. Nuclear Regulatory Commission, EIB/RES, M.S. 302 NL/S, Washington, DC 20555.

Ken J. KarisAllen[1] *and Jack Morrison*[2]

Mechanical Reduction of Inertially Generated Effects in Single-Edge Notched Bend (SENB) Specimens Subjected to Impact Loading

REFERENCE: KarisAllen, K. J. and Morrison, J., **"Mechanical Reduction of Inertially Generated Effects in Single-Edge Notched Bend (SENB) Specimens Subjected to Impact Loading,"** *Rapid Load Fracture Testing, ASTM STP 1130*, Ravinder Chona and William R. Corwin, Eds., American Society for Testing and Materials, Philadelphia, 1991, pp. 76–91.

ABSTRACT: Two novel mechanical techniques are described for reducing the magnitude of the inertial component of dynamic load/time records while maintaining a high specimen velocity response. One approach uses a load-line plastic absorber to inhibit the oscillation from establishing itself; the second uses precompressed wing absorbers to reduce selectively the amplitude of the load oscillations commonly associated with inertial energy transfer. Both systems independently generate load-time records in which the magnitude of the oscillations is greatly reduced without a significant change in vibration frequency. The load-line also reduces the magnitude of the initial tup load spike commonly associated with impact testing. The data indicate that the use of these absorbers does not introduce any extraneous effects in the dynamic load/time records and that maximum damping can be obtained by combining the two techniques. The elastic portion of the dynamic and static load/displacement data also shows a good correlation.

KEYWORDS: fracture, dynamic loading, three-point bend specimen, mechanical damping, drop tower, impact testing

Drop-weight and pendulum machines have been used historically for qualitative assessment of the dynamic properties of materials [*1,2*]. Early attempts to quantify the information obtained highlighted the need to understand the loading interaction between the machine and the specimen [*3,4*]. It was determined that, particularly during the early portion of the impact test, load sensors mounted on the anvils or striker do not reflect actual crack-tip loading because of the vibrations present in the system [*5*]. Various specimen/system configurations and load transducer positions have been investigated [*4,6*]. It has been shown that if the machine/specimen contact stiffness ratio is sufficiently low, then the crack tip experiences vibration-free loading [*7*]. In addition, the magnitude of the initial load spike sensed by a striker-mounted transducer can be controlled by judicious selection of impact velocity and also by matching the acoustic impedances of the striker and specimen [*8*]. However, this demands a unique arrangement tailored to the material being tested.

It is generally accepted that the load oscillations result from the inertial energies transferred across the tup/specimen/anvil interfaces on initial impact. It has been shown that the initial motion of a bend specimen contacted by a free-falling striker assembly is not influenced by the presence of anvil supports [*9*]. In this scenario, the specimen rotates away from the anvils about

[1] Defence scientist, Defence Research Establishment Atlantic, P.O. Box 1012, Darthmouth, Nova Scotia, Canada, B24 3Z7.

[2] Defence scientist, Defence Research Establishment Pacific, FMO Victoria, British Columbia, Canada, V0S 1B0.

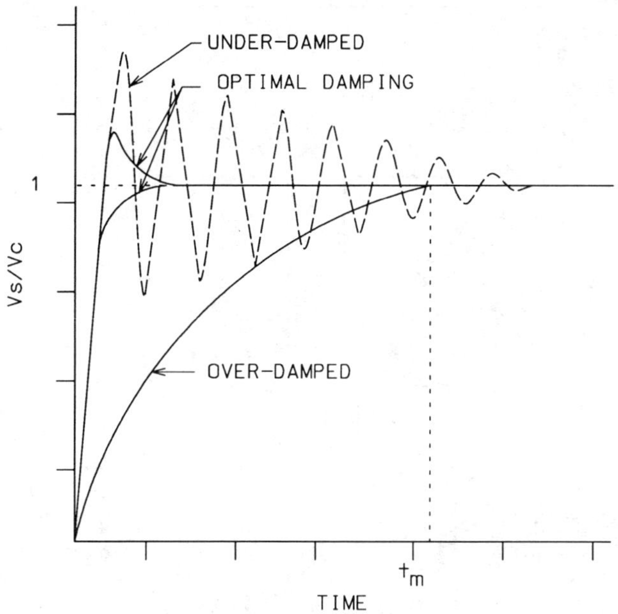

FIG. 1—*Schematic representation of different velocity-time curves.*

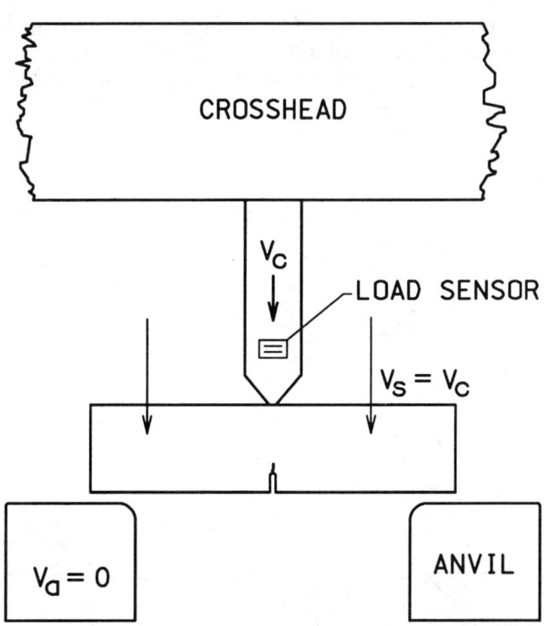

FIG. 2—*Tup/specimen/anvil velocities before impact.*

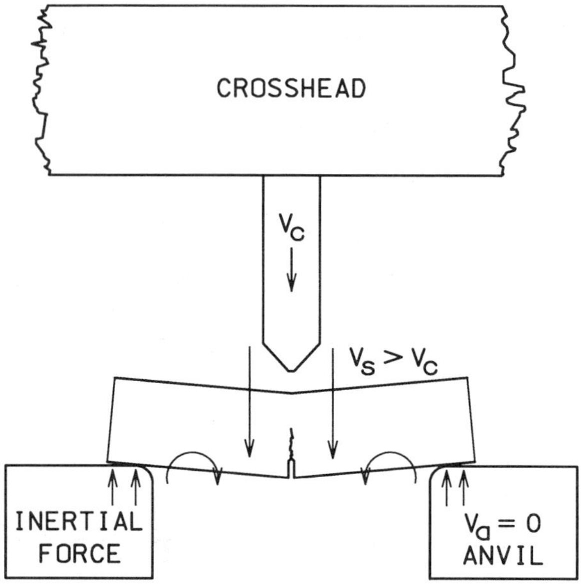

FIG. 3—*Tup/specimen velocities on impact.*

nodes located between the center and ends of the specimen, while the striker remains in contact with the specimen. The striker continues to deflect the specimen, establishing a three-point bending loading component. If the load sensor is an integral part of the striker, then the resulting force/time record reflects these relative motions as a free-body oscillation superimposed on a monotonic loading. Sensors located close to the vibrational nodes tend to increase the relative

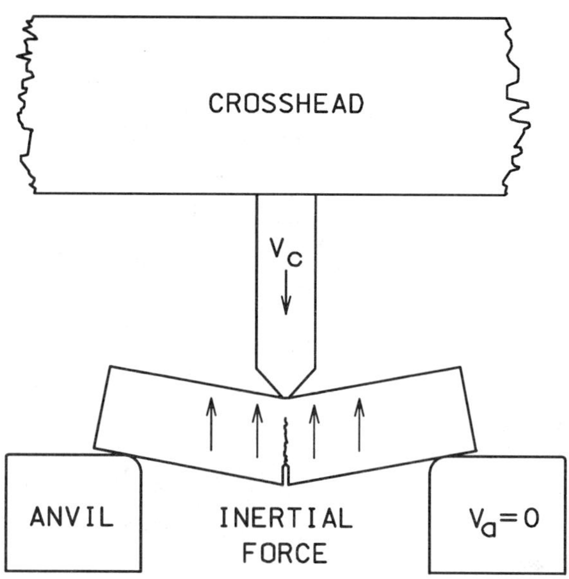

FIG. 4—*Tup/specimen/anvil velocities at specimen/tup recontacting.*

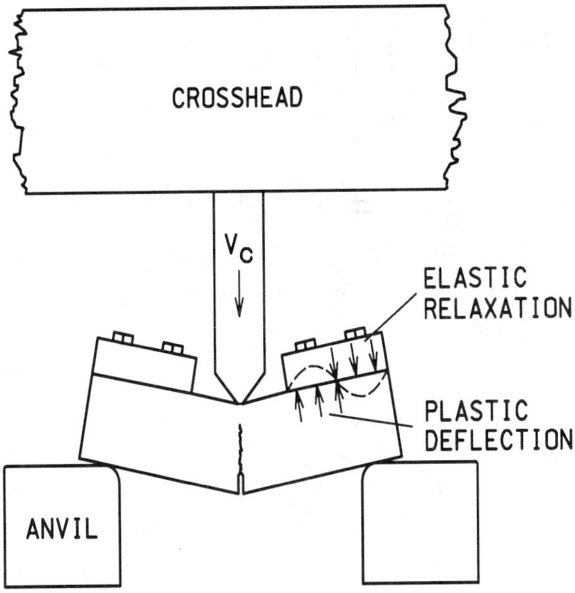

FIG. 5—*Schematic representation of wing absorber response to specimen oscillation deflection.*

magnitude of the three-point bending component of the signal [*9,10*]. The resulting force/time data can be smoothed numerically to make it possible to analyze the test data. However, the resulting record does not give an accurate representation of the stress/time state at the crack tip, and it also carries the risk of skewed inflections and discontinuities in the load data as a result of excessive smoothing. It has been shown that a linear-elastic toughness can be obtained from a load trace containing oscillations provided that the critical fracture event does not occur before an elapsed time of approximately three times the period of the load oscillations [*11*]. This approach cannot be applied to many low-toughness materials.

Perhaps the most promising method of improving the processing of impact data, while providing an accurate representation of the crack-tip stress state, is through the mechanical damping of the inertial component of the test. Figure 1 compares schematically the ratio of the specimen mid-section velocity V_s to the crosshead velocity V_c for several configurations. The underdamped case is characteristic of a typical impact test that relies on the natural damping characteristics of the specimen material to remove the inertial oscillation. During this type of event, the striker may lose contact with the specimen several times during the test [*9,10*], thus degrading the usefulness of the striker data. It has been shown that, if a soft metal absorber is inserted between the specimen and striker, then the oscillatory component of the record can be effectively removed [*12*]. This case corresponds to the overdamped velocity/time trace in Fig. 1. Although providing a marked improvement over the underdamped case, the velocity response of the specimen has been degraded (the specimen reaches the crosshead velocity only after *tm* (time to maximum load)). To provide the desired dynamic impulse at the crack tip, the velocity response of the specimen must be increased toward that represented by the optimally damped velocity/time profiles. This should represent the best compromise between response and damping. It is the goal of the research carried out at the Defence Research Establishment Pacific to explore novel mechanical methods of reducing the magnitude of the inertial oscillation component of single edge-notched bend (SENB) specimens while maintaining the required dynamic response.

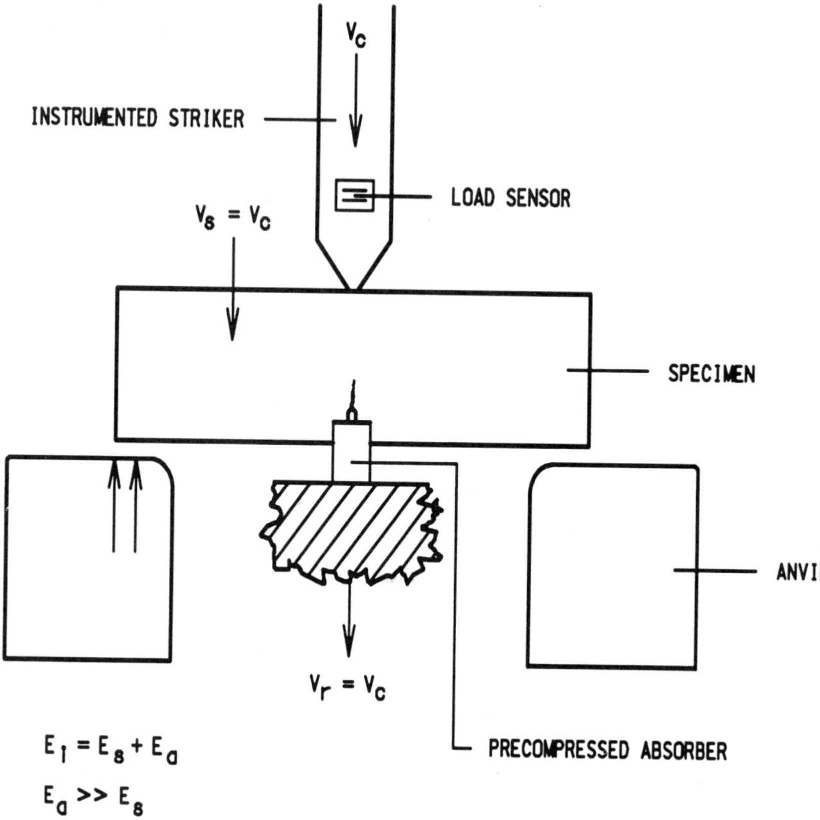

FIG. 6—*Schematic representation of the load-line absorber configuration.*

Experimental Program

Background

If a standard drop tower is modified such that the specimen is attached in a nonrigid manner to the crosshead (Fig. 2), then the inertial effects are generated at the anvil/specimen interfaces. Because the specimen behaves as a nonrigid body, the resulting specimen motion will have translational and rotational components, the magnitudes being a function of the specimen bending stiffness. Applying the arguments of Böhme et al. [*9*], the overall effect is the initial loss of contact between the specimen and the striker (Fig. 3). The initial impact and the secondary recontacting impact (Fig. 4) combine to establish a specimen oscillation, the crack tip experiencing an underdamped condition. The oscillation has deflection components in both forward and reverse directions with respect to the specimen and crosshead motion. The reverse deflection component can be removed by attaching to the upper surface on each side of the specimen metal wing absorbers precompressed to yield (Fig. 5). In this configuration, the relative deflection in the same direction as the crosshead motion produces elastic relaxation in the absorber, whereas the opposing deflection produces plastic deflection in the absorber. The oscillation can be removed selectively with negligible effect on the three-point bending component. Although it

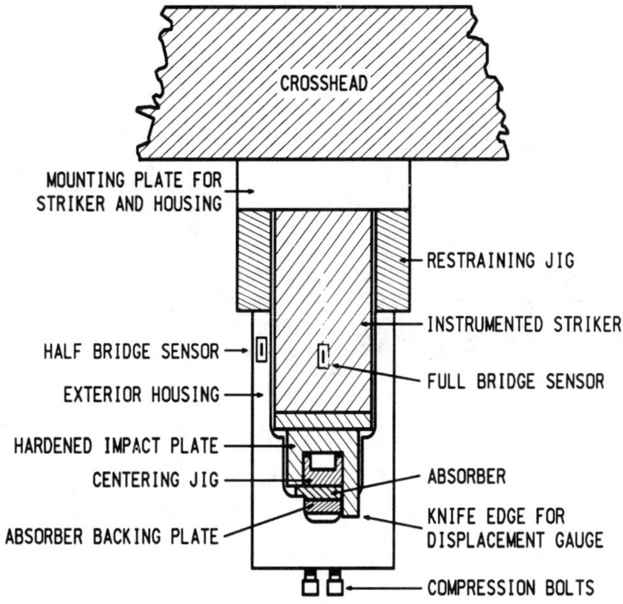

FIG. 7—*Front view of modified striker with exterior restraining housing.*

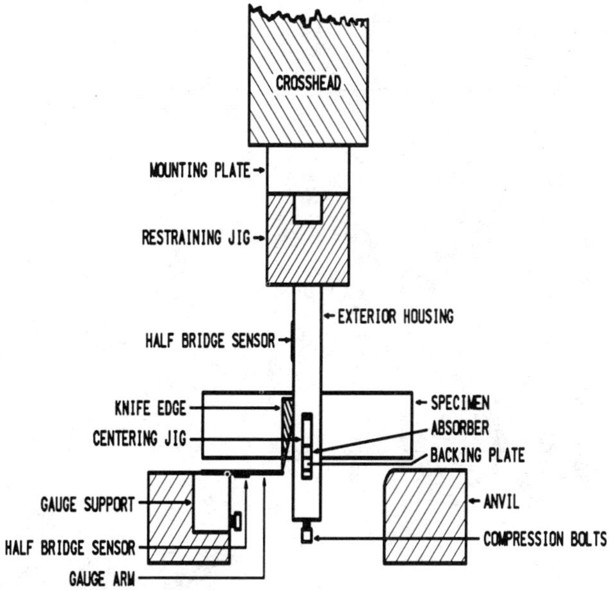

FIG. 8—*Side view of striker/specimen/anvil configuration showing displacement gauge positioning.*

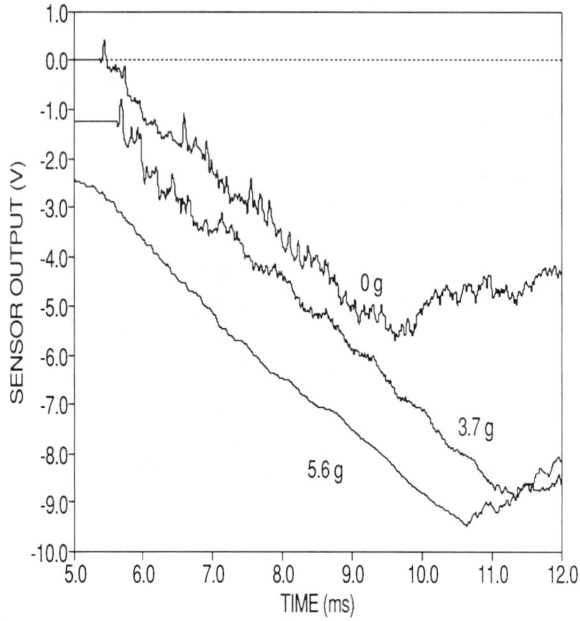

FIG. 9—*Offset voltage-time signals for cantilever arm displacement gauge impacted at 1 m/s showing the improvement in vibrational noise as the additional load point mass is increased from 0 to 3.7 to 5.6 g (no specimen).*

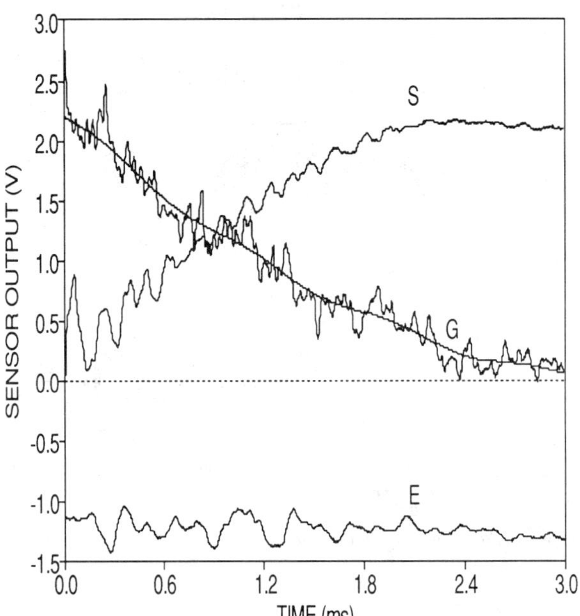

FIG. 10—*Typical gauge displacement* (G), *striker load* (S), *and exterior housing load signals* (E) *for a specimen impacted at 1 m/s.*

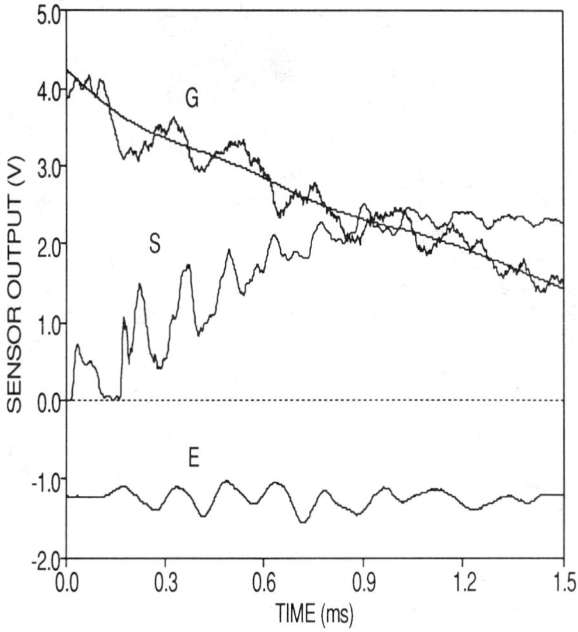

FIG. 11—*Typical gauge displacement* (G), *striker load* (S), *and exterior housing load signals* (E) *for a specimen impacted at 2 m/s.*

provides a suitable means of damping the oscillation, this arrangement has the disadvantage of there being an initial temporary loss of contact between the striker and specimen.

An alternative arrangement is to attach the specimen semi-rigidly to the crosshead by compressing a metal absorber to yield against the lower surface of the specimen along the load line (Fig. 6). In this case, the deflection energies along the load line resulting from the initial rotation of the specimen away from the striker are absorbed elastically by the specimen deflection and also plastically through absorber deflection. The ratio of specimen to absorber energy can be manipulated by controlling the plastic flow properties of the absorber material. Increases in plastic energy absorption tend to push the velocity profile of the specimen from the underdamped case toward the optimally damped velocity profile. This research explores these two methods of reducing the magnitude of the oscillatory component of the impact load history while maintaining a high initial velocity response.

Procedure

Several modifications were incorporated into a standard drop tower to investigate the ideas presented above (Figs. 7 and 8). The instrumented load striker was modified by the addition of an exterior housing, which acted as a platform for the various forms of restraint required. This housing was fixed to the crosshead using restraining blocks and four bolts threaded through the mounting block and into the crosshead. A slot was machined into the head of the housing and used to align any load-line absorber. The magnitudes of deflection and load applied to the absorber were controlled by three threaded bolts tapped through the head and into the slot. The load applied to the bolts was monitored by a half bridge transducer located on the arm of the

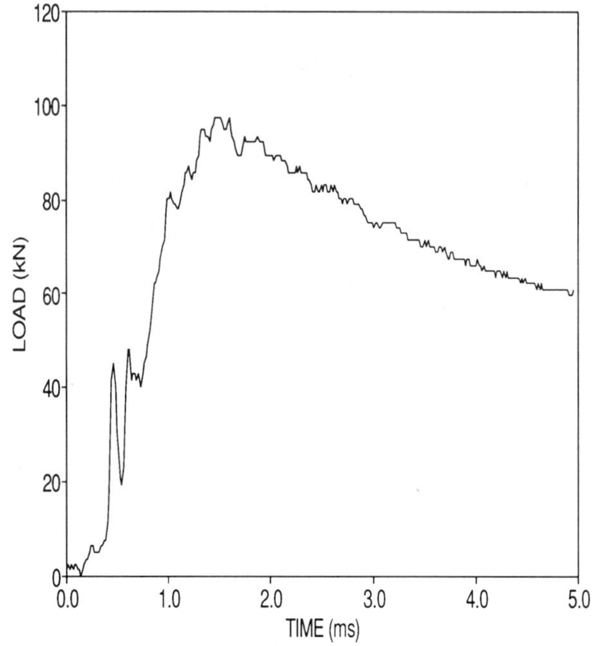

FIG. 12—*Load-time record for a specimen with wing absorber impacted at 2 m/s.*

exterior housing. Positioning of the specimen was achieved by an in-situ jig that centers the specimen crack plane with respect to the instrumented load cell using the machined notch. The Brinelling effect of the load cell on the specimen was minimized by machining a flat surface on the impact plate. Dynamic displacement was monitored using a 0.6-mm-thick spring steel cantilever-style arm coupled with a half bridge transducer located on the lower base of the arm. The deflection was measured by a contact gauge, the arm being displaced by a knife-edged block attached directly to the specimen/load cell contact surface. Total gauge deflection was limited by sets of 200- by 200-mm steel blocks of various heights, which were used to arrest the crosshead at predetermined displacement levels. All load and displacement sensors were calibrated statically using a servohydraulic materials testing system.

SENB specimens (25.4 by 50.8 by 254.0 mm) produced in accordance with ASTM Test for Plane-Strain Fracture Toughness of Metallic Materials (E 399) were machined from HY130 steel and precracked to nominal 0.55 a/W. Testing was carried out by placing the specimen against the impact plate, positioning the centering jig into the notch and preloading using the compression bolts and various forms of absorber. Load-line absorber material varied depending on the level of restraint required for a particular test. In increasing order of restraint, absorber materials included packing foam (no restraint), thermoset poly(methyl methacrylate) (Lucite), and annealed aluminum. Wing absorbers were machined from commercially pure lead. The 225-kg crosshead was raised to a predetermined height, and the specimen was impacted on the flat hardened anvils at velocities of approximately 1 and 2 m/s, deflecting the specimen in three-point bending. The anvil span was set at four times the specimen width of 50.8 mm in accordance with ASTM Test E 399. The three transducer signals were amplified using high-speed conditioners, and the signals were routed to a 12-bit 256K per channel oscilloscope where

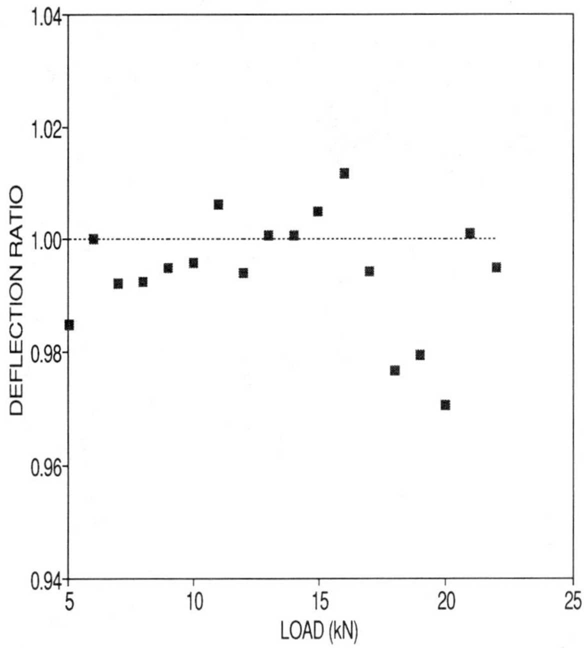

FIG. 13—*Ratio of static deflections as a function of load for a specimen with wing absorbers attached compared to a standard specimen.*

they were digitized and stored. Dynamic records were subsequently bussed to a microprocessor for analysis.

Results and Discussion

It was important to establish that the displacement gauge was capable of accurately monitoring the dynamic motion of the crosshead. Because of the large forces imposed on the crosshead by the stop blocks, a maximum of 1-m/s impact could be used for these tests when no specimen was present. A typical system response for the displacement gauge revealed a high-frequency component inherent in the gauge as a result of striker/displacement gauge impact (Fig. 9). The period and amplitude of this oscillation were improved by the addition of a small mass to the impact point on the cantilever, which effectively changes the vibrational response of the gauge. The quality of the trace also dramatically improved as the additional mass at the load point was increased from 0 to 3.7 to 5.6 g. Note, however, that when crosshead/arrestor impact occurred, in all three tests the level of vibration increased significantly. Several tests were conducted to compare dynamically measured voltages with statically measured voltages. The common displacement level chosen was the point where crosshead/arrestor impact occurred, because this point was easily obtained both dynamically and statically. The accuracy and reproducibility of such values agreed within 1 percent, the dynamic value being consistently higher. Because the gauge showed good linearity over a range in excess of 10 mm in this configuration, it was determined that this device should produce acceptable displacement records.

Figures 10 and 11 show typical displacement, striker load, and exterior housing load records

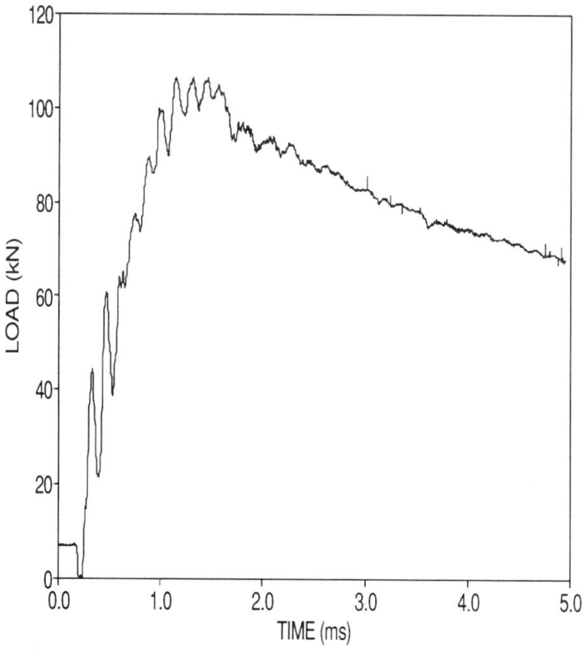

FIG. 14—*Load-time signal for a specimen with a Lucite load-line absorber impacted at 2 m/s.*

for 1- and 2-m/s impacts when no specimen restraint is present (that is, styrofoam absorber). At 1 m/s, the initial spike on the load/time record is quite pronounced with ringing continuing up to maximum load. Overall, however, even with the undamped oscillations, the signals are judged to be marginally acceptable for data analysis. The high-frequency vibrational component is again observed in the displacement signal on specimen impact. There is also a second lower-frequency oscillation, which roughly follows the load oscillations, possibly resulting from inertial energies being transferred during specimen/anvil impact. In the modified drop tower, with the specimen attached to the crosshead, the test is equivalent to a stroke-controlled impact. Consequently, no inflections or discontinuities in velocity are expected, provided that the inertial oscillations are removed. Under these conditions, a piece-wise low-order iterative polynomial curve fit (shown in Figs. 6*a* and 6*b*) or some type of digital filtering should effectively filter this high-frequency gauge oscillation with a minimum loss of specimen displacement data.

Comparing Fig. 6*a* to Fig. 6*b* shows a substantial increase in the amplitude of the load oscillation as the velocity is increased from 1 to 2 m/s without specimen constraint. The first peak results from specimen misalignment (one end of the specimen impacting the anvil before the other). This slight misalignment was observed frequently in tests without load-line constraint, where the specimen is held only loosely in the housing. The second oscillation occurs while full contact is being established and results in a doubling of the amplitude at the higher velocity, in agreement with published observations [*8*]. The inertial oscillations contribute a significant portion of the overall load sensed by the striker before maximum load. For this configuration and velocity combination, the striker loads and displacement sensors probably do not provide an accurate specimen history as a result of the loss of contact between the specimen and striker at various times during the test. Studies have shown that critical elastic events occurring after an

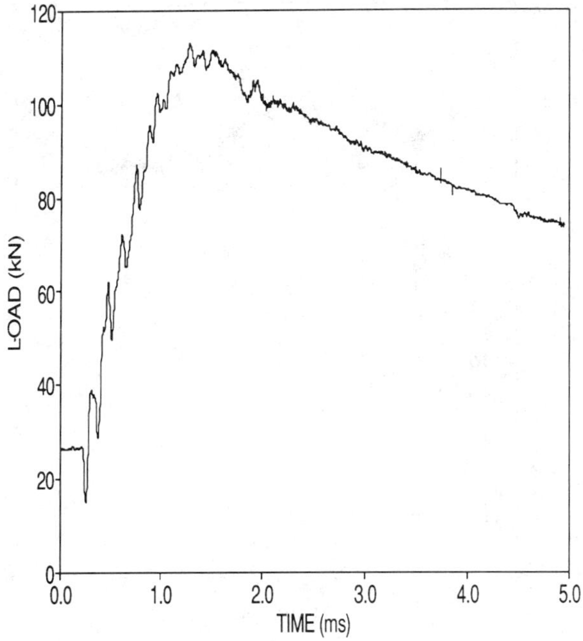

FIG. 15—*Load-time signal for a specimen with an annealed aluminum load-line absorber impacted at 2 m/s.*

elapsed time greater than three periods of oscillation can be quantified accurately using this type of signal [*11*]. However, single-specimen elastic-plastic analyses tend to be very sensitive to the slope of the load/deflection trace [*10,13*]. Because 2 m/s was the largest initial velocity obtainable for these specimens, without running the risk of damage to the drop tower because of the massive stop blocks, this velocity was used to demonstrate the damping characteristics of the wing and load-line absorbers (at higher velocities, commercial drop towers require modification to withstand the very high shear forces on the crosshead bolts). In both the 1- and 2-m/s tests, the period of oscillation agrees closely with that predicted for a Mode 2 vibration [*11*].

A disappointing feature of these preliminary tests was the response of the transducer fixed to the arm of the exterior housing. The fact that there was no preload on the fixture and no interaction between specimen and housing indicates that any load response was simply due to the change of inertia in the fixture combined with a lack of stiffness in the housing arms. The presence of such a large oscillation would make any preloaded absorber data from this gauge suspect, and for this reason it was not used for remaining tests. Instead, the absorber loads were inferred indirectly from the striker load cell signals.

Figure 12 is the load/time record from a specimen for which precompressed lead wing absorbers were used to absorb the oscillation selectively. As expected, the recontacting initial inertial spike is still present and is even larger in magnitude as a result of the increase in specimen mass resulting from absorber attachment. The magnitude of the subsequent oscillations is reduced dramatically with respect to the undamped case to the point of marginal acceptability with respect to most fracture analysis techniques. The natural period of oscillation for this geometry is increased slightly over that of the undamped tests. The attachment holes for the wing absorbers were kept to the minimum usable depth and purposely located remote from the crack tip so as to

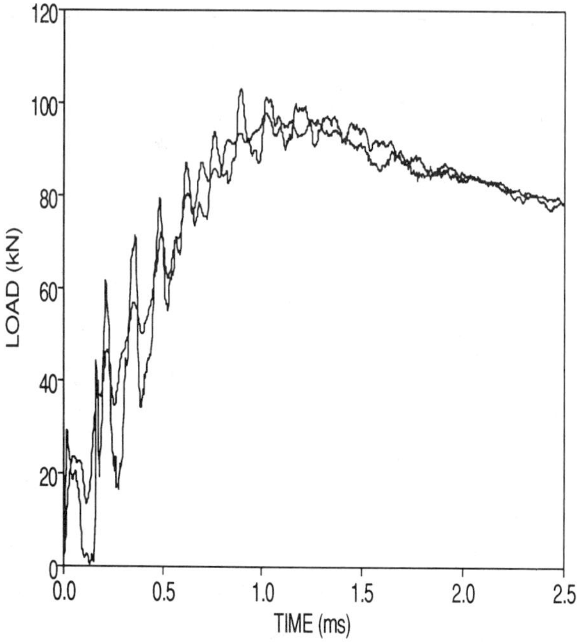

FIG. 16—*Load-time record comparison of unconstrained and aluminum load-line absorber test. The minimum observed load has been subtracted from the aluminum signal.*

have minimal effect on the stress fields generated at this point. However, globally, the increased period reflects a decrease in the effective stiffness or an increase in the effective mass of the specimen [5,11]. Changes in the stiffness/geometry relationship could change the $f(a/w)$ portion of the K solution for this specimen. This was investigated by statically measuring the compliance of a specimen and subsequently measuring the compliance for the same specimen modified by the precompressed absorbers. This comparison (Fig. 13) indicates that the addition of the wing absorbers has negligible effect on the effective modulus of the specimen and the increase in oscillation period is probably a result of the increase in effective mass of this configuration.

Figures 14 and 15 are the load-time records obtained when Lucite and annealed aluminum, respectively, were precompressed to yield on the bottom surface of the specimen along the load line. It is noted that, as the yield strength of the material increases, the amplitude of the initial oscillation decreases with respect to the unrestrained case. In both cases, the period of oscillation is comparable to the undamped test. Although the Lucite absorber significantly reduces the magnitude of the initial spike, oscillations still persist up to maximum load. As such, this signal would require further processing to be usable. Alternatively, some additional damping capability is necessary. On the other hand, the aluminum absorber produces a load-time record with greatly reduced vibration amplitudes. It is, however, important to note that, although in the case of the Lucite the energy transferred to the absorber in the form of plastic deflection removed the preload completely during initial impact, only partial removal occurred in the case of the aluminum absorber because of its higher flow stress. Figure 16 compares the load/time record of the unrestrained test to the aluminum absorber test when the minimum observed absorber load as indicated by the striker data is subtracted from the trace. In this case, the minimum load is taken to approximate the point of maximum plastic deflection in the aluminum absorber. The correla-

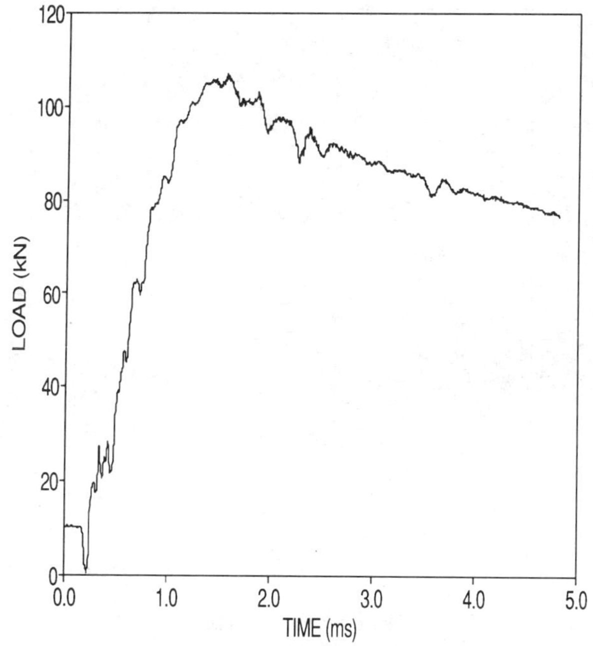

FIG. 17—*Load-time record for a specimen with both wing and Lucite load-line absorbers impacted at 2 m/s.*

tion between the records is excellent. It is not clear to what extent the residual absorber compression influences the hydrostatic component of the crack-tip stress fields. This will be the subject of a further study but is beyond the scope of this investigation. For the purposes of this paper, only the Lucite records, in which the absorber load is completely removed during the initial impact, are given further consideration.

Because the wing absorbers and load-line absorbers reduce the oscillations by different mechanisms, the damping capabilities should be complementary. Figure 17 represents the load/time signal when a Lucite load-line absorber is used in combination with lead wing absorbers. As expected, the combined effect of the two damping techniques reduces the load oscillations throughout the test. Again, the introduction of the lead wing absorbers to the system slightly increased the period of the oscillation. Because the specimen material stiffness is not very rate sensitive, statically measured data should provide a check of the accuracy of the dynamic data and a check of any stiffness changes due to the testing configuration. An excellent agreement is found between static and dynamic stiffness (Fig. 18). A good correlation is found when the static elastic stiffness of the same specimen is compared to that determined from dynamically measured load-line displacement. This correlation again indicates that the increase in the period of oscillation results from the increase in the effective mass of the specimen. Further testing is being carried out to establish the data reproducibility and to determine the correlation between dynamic load data derived from both the tup and from gages mounted in the vicinity of the crack tip.

Conclusions

Two mechanical configurations were tested for their ability to suppress the load oscillations for SENB specimens subjected to impact loading rates. Wing absorbers effectively dampened

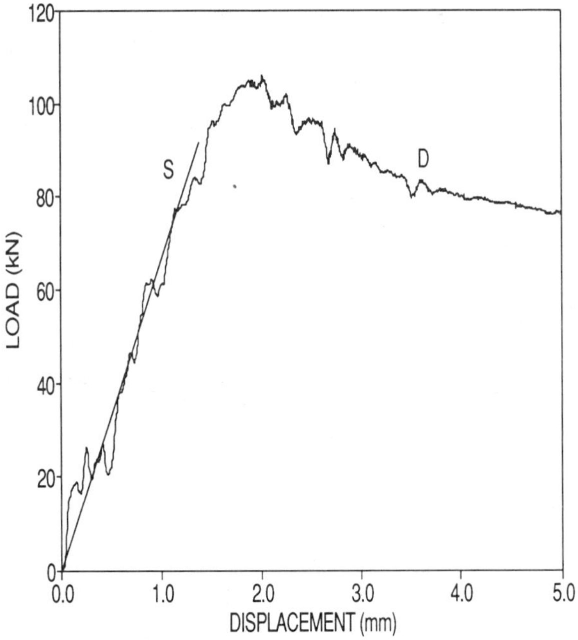

FIG. 18—*Load-displacement record for the test shown in Fig. 11 comparing the dynamic signal* (D) *to the previously measured static stiffness* (S).

oscillations, although a recontacting load spike was still present. The load-line absorber significantly reduced the initial spike (the reduction increasing with absorber yield strength). However, further work will have to be carried out on the effect of residual absorber compression before this method becomes a viable technique. The two absorber techniques in combination effectively reduced the load oscillation over the entire test as a result of their complementary damping capabilities. For the material tested, the dynamic elastic load-displacement data correlated well with statically measured data. Further work is being carried out to determine the reproducibility of the system. The displacement gauge is being modified with a mechanical damping system to eliminate the need for numerical smoothing.

References

[*1*] Lange, E. A. and Loss, F. J., in *Impact Testing of Metals, ASTM STP 466*, American Society for Testing and Materials, Philadelphia, 1970, pp. 241–258.

[*2*] Fahey, N. H., in *Impact Testing of Metals, ASTM STP 466*, American Society for Testing and Materials, Philadelphia, 1970, pp. 76–88.

[*3*] Turner, C. E., in *Impact Testing of Metals, ASTM STP 466*, American Society for Testing and Materials, Philadelphia, 1970, pp. 93–112.

[*4*] Trudeau, L. P., "Dynamic Toughness—Its Relevance and Measurement," Research Report R 275, Department of Energy, Mines, and Resources, Mines Branch, Canada, Oct. 1974.

[*5*] Venzi, S., Priest, A. H., and Wylie, R. D., in *Impact Testing of Metals, ASTM STP 466*, American Society for Testing and Materials, Philadelphia 1970, pp. 165–180.

[*6*] Glover, A. P., Mucci, P. D. R., and Radon, J. C., "Time Measurement in the Fracture Toughness Studies," *Engineering Fracture Mechanics*, Vol. 2, 1970, pp. 165–167.

[7] Turner, C. E., Culver, L. E., Radon, J. C., and Kennish, P., "An Analysis of the Notched Bar Impact Test with Special Reference to the Determination of Fracture Toughness," presented at Institute of Mechanical Engineers Conference, London, 3–5 May 1971, Paper C6/71, pp. 38–47.
[8] Saxton, H. J., Ireland, D. R., and Server, W. L., in *Instrumented Impact Testing, ASTM STP 563*, American Society for Testing and Materials, Philadelphia, 1974, pp. 50–73.
[9] Böhme, W. and Kalthoff, J. F., "The Behaviour of Notched Bend Specimens in Impact Testing," *International Journal of Fracture*, Vol. 20, Dec. 1982, pp. R139–R143.
[10] Joyce, J. A. and Hackett, E. M., "Dynamic *J-R* Curve Testing of a High Strength Steel Using the Key Curve and Multispecimen Techniques," in *Fracture Mechanics: Seventeenth Volume, ASTM STP 905*, American Society for Testing and Materials, Philadelphia, 1986, pp. 741–774.
[11] Ireland, D. R., "Critical Review of Instrumented Impact Testing," presented at International Conference on Dynamic Fracture Toughness, London, July 1976, Paper 5, pp. 47–62.
[12] Joyce, J. A. and Hackett, E. M., "An Advanced Procedure for *J-R* Curve Testing Using a Drop Tower," in *Nonlinear Fracture Mechanics, ASTM STP 995*, American Society for Testing and Materials, Philadelphia, 1989, pp. 298–317.
[13] Herrera, R. and Landes, J. D., "A Direct *J-R* Curve Analysis of Fracture Toughness Test," *Journal of Testing and Evaluation*, Vol. 16, No. 5, Sept. 1988, pp. 427–449.

Wolfgang Böhme[1]

Fracture Resistance of a Pressure Vessel Steel Under Impact Loading Conditions

REFERENCE: Böhme, W., **"Fracture Resistance of a Pressure Vessel Steel Under Impact Loading Conditions,"** *Rapid Load Fracture Testing, ASTM STP 1130*, Ravinder Chona and William R. Corwin, Eds., American Society for Testing and Materials, Philadelphia, 1992, pp. 92–103.

ABSTRACT: Impact experiments were performed with Charpy specimens and with large bend specimens made from a ferritic pressure vessel steel. All specimens were side-grooved and fatigue-precracked. Different amounts of ductile crack extension up to 16 mm were achieved by varying the available impact energy. The measured hammer load signals were evaluated to calculate *J* integral values.

The determined dynamic *J*-resistance curves of the bend specimens are significantly above the Charpy data. Examinations of the fracture surfaces and metallographic sectioning perpendicular to the crack front showed that behind the stretched zone the ductile main crack is accompanied by side cracks with a length up to about 3 mm for the large specimens, whereas such side cracks are hardly found in Charpy specimens. The additional energy needed for the formation of side cracks might explain higher *J*-resistance curves for larger specimens.

Further results indicate that with increasing loading rate and specimen size the tendency to develop side cracks increases in correlation with higher *J*-resistance curves. Finally, an attempt is made to discuss the relevance of this effect for the behavior of components.

KEYWORDS: ductile fracture, *J*-integral, crack resistance curve, multiple cracks, side cracks, crack branching, impact test, high loading rate

The fracture toughness of ductile materials can be quantified for static loading situations on the basis of the *J*-integral concept following standardized procedures, for example, ASTM Test for J_{Ic}, a Measure of Fracture Toughness (E 813) or ASTM Test for Determining *J-R* Curves (E 1152). An increasing effort has been undertaken into the determination of *J*-resistance curves under higher rates of loading. Such so-called dynamic *J-R* curves were determined by using different experimental techniques and evaluation procedures (see Refs *1* through *7*). The obtained results mainly indicate that dynamic *J-R* curves are higher than static ones.

The aim of this investigation was the characterization of a pressure vessel steel under high rates of loading. Therefore, impact tests with impact velocities of several meters per second were performed. The achieved loading rates were five to six orders of magnitude above usual static rates and can be considered as an upper limit for loading rates expected during postulated emergency and faulted conditions in a pressurized water reactor.

Ferritic pressure vessel steel with the German designation 22 NiMoCr 3 7 (≙ ASTM A 508 Cl 2) was investigated. The tests were performed at temperatures in the upper-shelf regime of the Charpy energy of about 200 J. Loading rate effects are quantified by comparison with results of static tests. Specimens of different sizes were tested, and observed size effects are discussed. Unexpected multiple crack formations focused the evaluations on the fracture processes during

[1] Fraunhofer-Institut für Werkstoffmechanik, Wöhlerstrasse 11, D-7800 Freiburg, Germany.

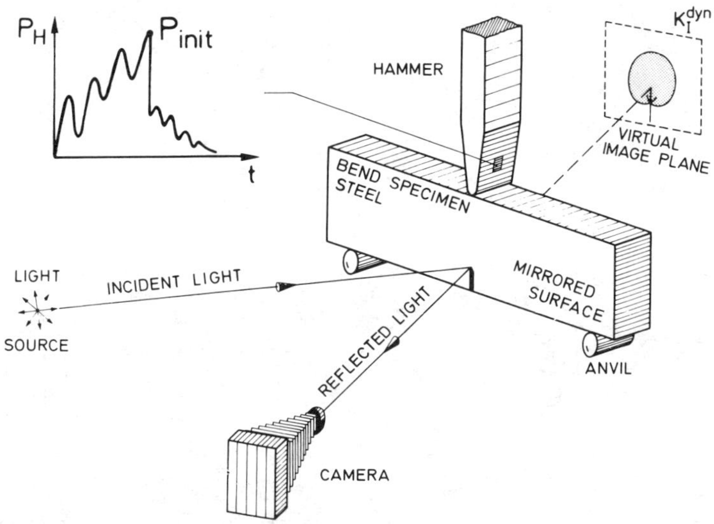

FIG. 1—*Experimental setup (schematic).*

ductile crack initiation and extension. The results are summarized here, and more details are given in Ref *8*.

Procedure

From a heat of the steel 22 NiMoCr 3 7 (≙ ASTM A 508 Cl 2) relatively large bend specimens (SENB 90) with dimensions 495 by 90 by 45 mm were machined in the L-S orientation. The in-plane dimensions are 9 times, and the thickness is 4.5 times larger than for precracked Charpy specimens. The specimens were fatigued and side-grooved (20%). The initial crack lengths a_0 were 45.6 ± 1.4 mm. The SENB 90 specimens were loaded in three-point bending with a support distance $S = 4W = 360$ mm in a drop-weight tower (Fig. 1). The impact velocity was 6.77 m/s. To achieve different amounts of ductile crack extension in different tests, the available impact energy was varied by using different masses of the hammer (111 to 282 kg).

From parts of a broken specimen without plastic deformations Charpy specimens (SENB 10) were machined and provided with fatigue cracks and side grooves like the SENB 90 specimens. The initial crack lengths were $a_0 = 5.5 \pm 0.3$ mm. The Charpy specimens were loaded in a pendulum with a mass of the hammer of 20 kg. Different amounts of ductile crack extension were achieved by a variation of the impact velocities between 0.55 and 1.73 m/s.

The procedures to vary the mass (SENB 90) or the velocity (SENB 10) are limited energy methods. It has been demonstrated in Ref *7* that the results are comparable to results obtained, for example, by stop-block experiments.

In all experiments, the load, P, was measured at the tup of the impacting hammer by means of statically calibrated strain gages and recorded as a function of the time, t. The time-dependent displacement of the hammer, $s(t)$, is then calculated from $P(t)$ by the second law of motion:

$$s(t) = \int_0^t \left[v_0 - \frac{1}{M} \int_0^t P(t)dt \right] dt \quad (1)$$

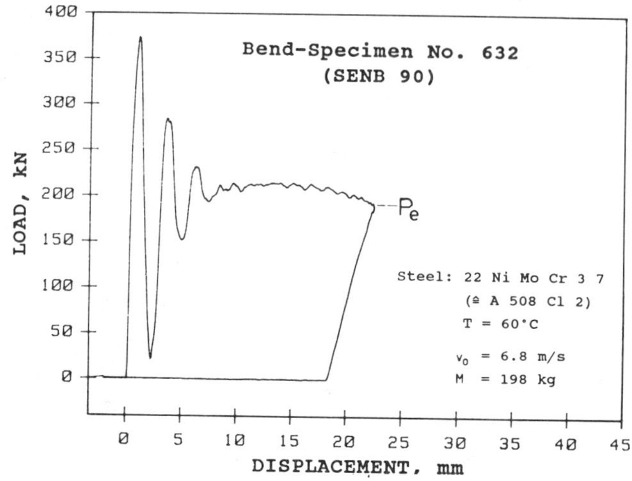

FIG. 2—*Load-displacement diagram* P(s) *for a SENB 90 specimen.*

where

v_0 = initial impact velocity and
M = mass of the impacting hammer.

Characteristic examples of the resulting load-displacement records $P(s)$ are given in Fig. 2 for the SENB 90 specimens and in Fig. 3 for the Charpy specimens. The load-line displacement, $\delta(t)$, of the specimens would be overestimated by the hammer displacement, $s(t)$, if loading devices of low stiffness are used. For checking the accuracy of Eq 1, therefore, the specimen deflection has been determined for the SENB 90 specimens by direct optical measurements using high-speed photography. The result was that $s(t)$ only moderately overestimates $\delta(t)$ by up to about 0.5 mm for SENB 90 specimens, which can be explained by the influence of the machine compliance ($C_m \approx$ 1.8 m/GN). For the SENB 10 specimens, the influence of the

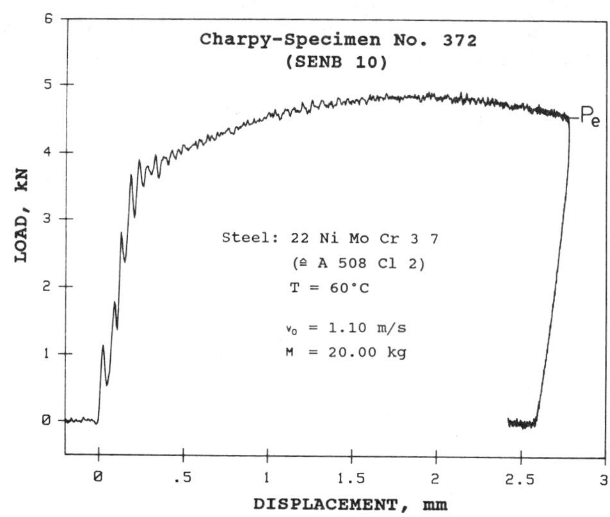

FIG. 3—*Load-displacement diagram* P(s) *for a Charpy specimen.*

machine compliance ($C_m \approx 11$ m/GN) was estimated to be only up to about 0.05 mm. For the displacements of interest (see Figs. 2 and 3), this influence is negligible. (It is of no influence on the following determination of J values, because the area of the load-displacement curve after unloading is considered as the absorbed energy, U_{pl}.)

Marked oscillations occur at the very beginning of the load signal with the first peak being the so-called inertia peak. These oscillations are caused by elastic wave propagation and vibration effects and are discussed elsewhere [*9–11*] in more detail. Because of damping effects like plastic deformations (especially near the impact position) these oscillations level out early. Therefore, the quasi-static evaluation performed here leads to reasonable accuracy after about three oscillations.

To gather information about the actual loading of the crack tip in addition to the external load record, almost all SENB 90 specimens were instrumented by strain gages. The foil gages (active length = 0.6 mm) were located at $\phi = 90°$ beside the crack tip at a distance of about 20 mm. Despite the oscillating external load in the very beginning of the impact event, the near crack-tip strain gage signals show an almost linear increase of the crack-tip loading.

This linear behavior is in agreement with earlier observations [*10,11*] and predictions [*11–13*] for this type of specimen. A quasi-static, nearly linear increase of the crack-tip loading can be expected in impact tests beyond the transition time, t_{dyn}, which has been established in Refs *11* and *12*:

$$t_{dyn} = 1.64\sqrt{C_s^*}\,\frac{W}{c_1} \tag{2}$$

where

C_s^* = dimensionless specimen compliance,
W = specimen width, and
c_1 = elastic wave propagation velocity (plane stress).

The calculated transition times, t_{dyn}, are 207 μs for SENB 90 and 23 μs for SENB 10. The corresponding displacements (calculated by impact velocity) are 1.4 and 0.02 mm, respectively. Therefore, by comparison with the displacements given in Figs. 2 and 3, it can be concluded that the crack-tip loading can be considered mainly as quasi-static without strong oscillations but, nevertheless, at high rates of loading.

From the slope of the elastic strain gage response, the initial crack-tip loading rates dK/dt are determined to be about 7×10^5 MPa $\cdot$ m$^{1/2}$/s for the SENB 90 tests (fixed impact velocity) and between 1.4×10^5 MPa $\cdot$ m$^{1/2}$/s and 3.2×10^5 MPa $\cdot$ m$^{1/2}$/s for the SENB 10 tests, depending on the impact velocity. These loading rates differ only by a factor of two to five and, therefore, can be considered as of the same order of magnitude, whereas usual static loading rates are about five decades lower.

The experiments were carried out at temperatures in the upper shelf of the Charpy energy (200 J), mainly at 60°C. A few SENB 90 specimens were also tested at room temperature and at 90°C. In some experiments at 60°C, a sudden change of the fracture mode into cleavage could be observed even after several millimetres of ductile crack extension [*8*]. In the following section, only the ductile crack extension is explored and evaluated. The results do not show a significant influence of the test temperature on the amount of ductile tearing so that the J-integral values and resistance curves are discussed independent of the temperature.

Results

Because for impact loading no standardized evaluation scheme is available to determine dynamic J-R curves, a multispecimen technique was chosen. The J-integral values are determined by ASTM Test E 813:

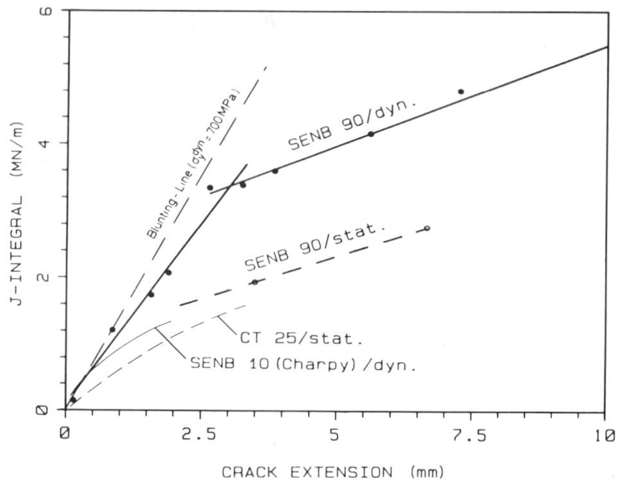

FIG. 4—*Static and dynamic* J-R *curves for specimens of different type and size (22 NiMoCr 3 7 ≙ A 508 Cl 2).*

$$J = J_{el} + \frac{2U_{pl}}{B_N(W - a_0)} \tag{3}$$

where the elastic part, J_{el}, is calculated following ASTM Test E 813 by using the load P_e (see Figs. 2 and 3) and

U_{pl} = area under load-displacement record (after unloading),
B_N = net thickness of the specimen, and
$W - a_0$ = original uncracked ligament.

The ductile crack extension, Δa, was determined after the test by optical inspection of the fracture surfaces.

Dynamic J values versus the respective amounts of ductile crack extension Δa are given in Fig. 4 for the SENB 90 and Charpy specimens. The linear shape of the *J-R* curve for the SENB 90 specimens for $\Delta a > 3$ mm is confirmed through further data points up to $\Delta a \approx 16$ mm. The *J-R* curve for the Charpy specimens results from ten data points with little scatter. Just for orientation, an estimated blunting line is additionally sketched in Fig. 4. It is calculated by $J = 2\ \sigma_y\ \Delta a$, where σ_y is the effective yield strength (average of the 0.2% offset tensile yield strength and the ultimate tensile strength). In static tension tests, the value $\sigma_y^{stat} = 560$ MPa was determined. For taking into account the increase of yield strength with increasing loading rate, the ASTM Test for Plane-Strain Fracture Toughness of Metallic Materials (E 399) was applied and an increase of about 25% was estimated. Therefore, the value $\sigma_y^{dyn} = 700$ MPa was used to calculate the given blunting line. The Charpy data yield J values up to about 1 MN/m, whereas the results from the SENB 90 specimens are significantly higher (see Fig. 4). The J values exceed the maximum J capacity, J_{max}, of the specimens as proposed in the standard after only small amounts of crack extension (Charpy specimen: $J_{max} = 0.35$ MN/m; SENB 90 specimen: $J_{max} = 1.58$ MN/m). This situation will not be improved by more detailed evaluations that, for example, consider crack-length corrections. Because better evaluation methods are not available, further discussion of the results is based on these J values.

The Charpy data are in agreement with comparable results for A 533 B and A 508 steels reported by Curr et al. [*3*] and Kobayashi et al. [*4*]. On the other hand, the SENB 90 data are very high. At first glance, the SENB 90 data seem to follow the estimated blunting line up to $\Delta a \approx 3$ mm. However, this steep increase of the data cannot correspond with blunting, because the end of blunting and beginning of crack extension occurs as early as $\Delta a \approx 0.1$ mm. Therefore, the initial steep increase is correlated with a significant amount of ductile crack extension. Only at very high J values of about 3.3 MN/m and $\Delta a \approx 3$ mm the slope of the curve changes remarkably, leading to a less steep increase of the resistance curve, however, at very high J values. This change in slope could be identified as a change in failure mechanism as discussed below.

For comparison, Fig. 4 shows further results of the same heat from static tests with SENB 90 specimens and with CT-25 specimens.[2] Two results from static SENB 90 tests define a slope that matches the slope of the dynamic curve for the SENB 90 specimens for large crack extensions. This points to the fact that the tearing modulus or the crack propagation energy will not be significantly different between static and dynamic loading.

Care has to be taken in comparing the absolute values, because they have been determined in a formal manner even outside the validity range. Numerical calculations performed by Thi Vo and Stöckl [*14*] for the impacted SENB 90 specimens indicate that the given J values have to be reduced by about 15%. This is due to energy consumption, for example, by plastic deformations and friction at the supports, which does not contribute to the crack-tip loading. Nevertheless, the static data are lower than the dynamic ones. This is apparently caused by the fact that for static loading the transition from the steeper to the flatter part of the curve occurs at smaller amounts of crack extension than for dynamic loading. The static curve intersects with the blunting line at a crack extension of about 1 mm as opposed to about 3 mm for the impact tests (see also next section).

The static J-R curve from the CT-25 compact specimens forms the lower bound to all other curves and is well in the range of other published data. The impacted SENB 90 specimens, however, yield data that are significantly higher (even if they are reduced by 15% as suggested by the numerical analysis) than comparable published results (see Refs *1* through *7*). The following analyses of the fracture surfaces contribute to an explanation of these high J values and the unusual kink of the J-R curve at $\Delta a \approx 3$ mm.

Fracture Analysis

Examination of the fracture surfaces by scanning electron microscopy revealed no indications for a change of the ductile fracture mode after a crack extension of $\Delta a \approx 3$ mm, which could explain the sudden drop in the slope of the resistance curve. Instead, at significant smaller crack extensions of $\Delta a \approx 0.3$ mm, crevices were observed running parallel to the crack front. By metallographic sectionings perpendicular to the crack front, macroscopic side cracks up to a length of 3 mm were detected below the ductile main crack, as demonstrated in Fig. 5.

This leads to the conclusion that in the beginning of the fracture process after crack-tip blunting at least two cracks were formed and extended. It might be expected that these two cracks consume more energy for propagation than a single crack, because, for example, for a given displacement the crack-tip opening for either of two (or more) nearby cracks will be lower than for a single one. This would qualitatively explain the initially steeper J-R curve if multiple cracks are present, and the subsequent smaller slope of the curve if only a single main crack is

[2] B. Voss, D. Sun, and L. Falk, "Static J-R Curves for SENB-90 and CT-25 Specimens for a Pressure Vessel Steel," unpublished results and private communication, IWM, Freiburg, 1988.

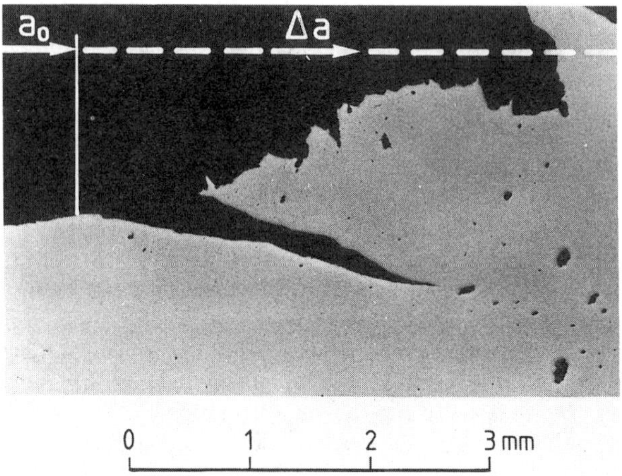

FIG. 5—*Polished section of a dynamically loaded SENB 90 specimen.*

propagating. In addition, the lengths of the observed side cracks agree well with the amount of crack extension where the change of slope occurs in the resistance curve.

It is furthermore remarkable that the main crack has a ductile fracture surface with dimples, whereas the side crack appears smooth (see Fig. 5), which might be an indication for shear failure along slip lines. This is confirmed also by microscopic investigations of the fracture surfaces as demonstrated in Fig. 6.

In one case, the crack-tip region of an impacted specimen was prepared by a special slicing technique (Fig. 7). Depending on the through-the-thickness position of the slice, two- and even

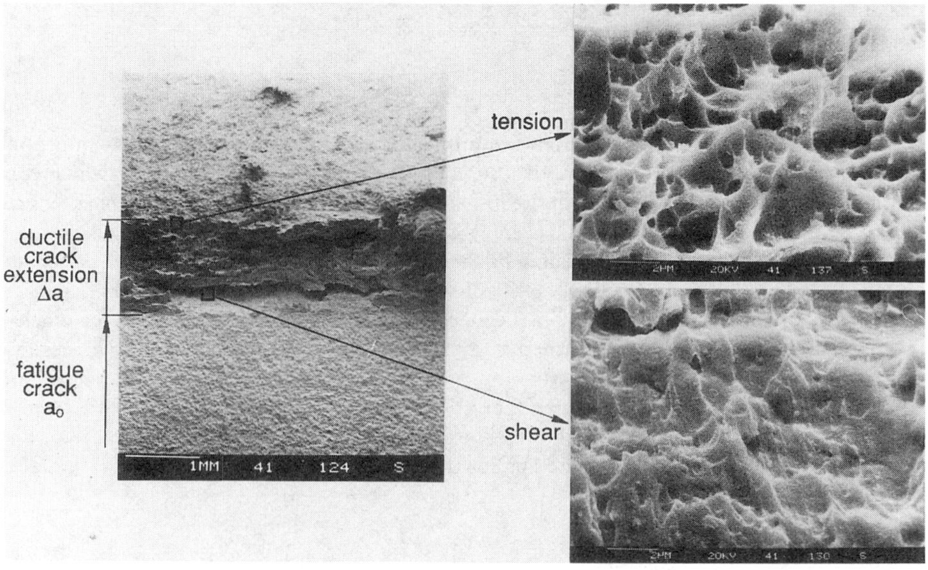

FIG. 6—*Different microscopic fracture appearance of the side crack and the main crack.*

FIG. 7—*Preparation of the crack-tip region by slicing.*

three-crack systems were found, and examples are given in Fig. 8. These multiple cracks occur just in front of the original fatigue crack. They can be observed across almost the whole thickness. Very close to the notches of the side grooves only single cracks are observable, which might be an indication that local constraint influences the development of side cracks. After a certain amount of crack extension, only one of these competing cracks continues to grow.

Thus, ductile crack initiation is accompanied here by macroscopic side cracks. The angle between the fracture surfaces of the main crack and the side cracks is about 45°. This indicates that side cracks can develop along slip lines at ±45° off the main crack. Therefore, two- or even three-crack systems are conceivable at initiation. Based on these results, some questions arise. One question concerns path independence of the J-integral if multiple cracks are present. It is another question if localized mixed-mode loading has to be taken into account and, therefore, J_I (tension) and J_{II} (shear) have to be considered.

Although a complete explanation of all details is not yet possible, the principal influence of

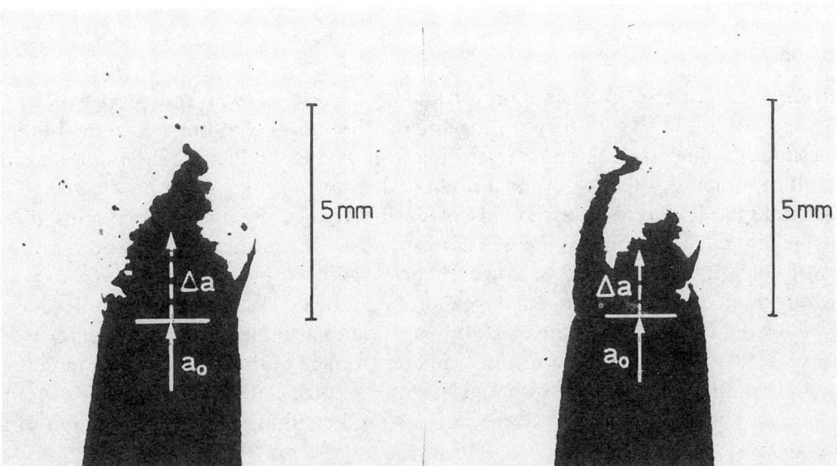

FIG. 8—*Examples of two- and three-crack systems.*

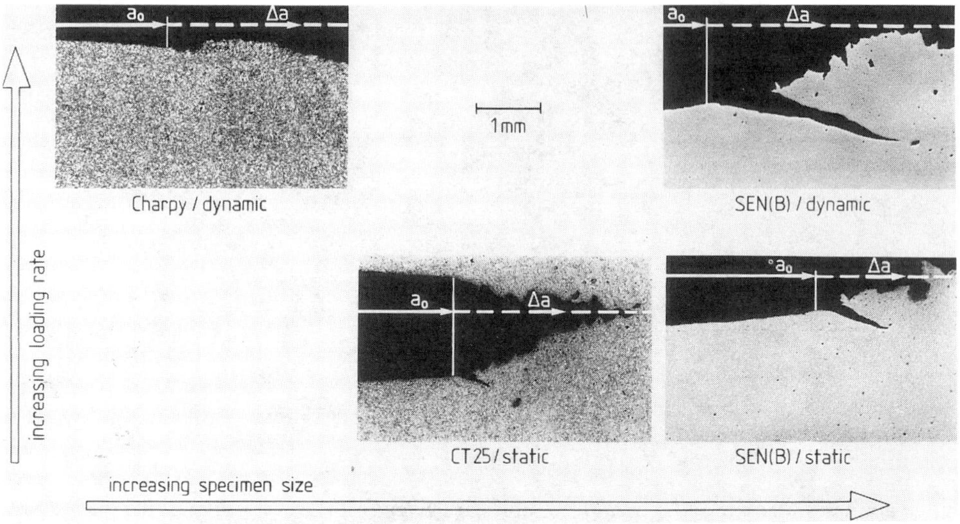

FIG. 9—*Polished sections of the crack-initiation region for different specimen and loading rates.*

loading rate and specimen size on this effect may be deduced at least qualitatively in Fig. 9. The side cracks are most pronounced for the dynamically loaded, large bend specimens (SENB 90), whereas side cracks hardly can be detected at the small Charpy specimens. Short side cracks are also observable at statically loaded CT 25 specimens, indicating that side-crack formation is not only a dynamic effect.

From Fig. 9 it can be concluded that, with increasing loading rate, and in particular with increasing specimen size, the tendency to form side cracks increases. At the same time, higher resistance curves with a significant change in slope are observed. This change occurs at amounts of crack extension that are in agreement with the size of the side cracks. These results are an indication that the tendency to develop side cracks is increasing with increasing stress triaxiality.

Discussion

Following the current understanding of ductile crack initiation and propagation, the first phase is characterized by crack-tip blunting and the formation of a stretched zone. In the second phase, a single ductile crack extends. Transition from Phase 1 to Phase 2 should be accompanied by a significant change in slope of the *J-R* curve defining a critical initiation value J_i.

This simple model has to be extended now in the light of the previously discussed results as sketched schematically in Fig. 10. Phase I corresponds to the former Phase 1 and is related to the process of crack-tip blunting. It is, however, now followed by a new Phase II, in which one propagating main crack is accompanied by additional side cracks. Only in Phase III (corresponding to the former Phase 2) does the crack propagation of a single ductile main crack take place.

Phases I, II, and III correspond to three different slopes of the resistance curve defining two kinks with two different critical J values. The value J_i corresponds to the well-known initiation value. The new value, J_i^{RL} is a critical value for the beginning of the propagation of a single ductile crack.

This fracture resistance value, J_i^{RL}, could be a material parameter that, however, depends strongly on the loading conditions.

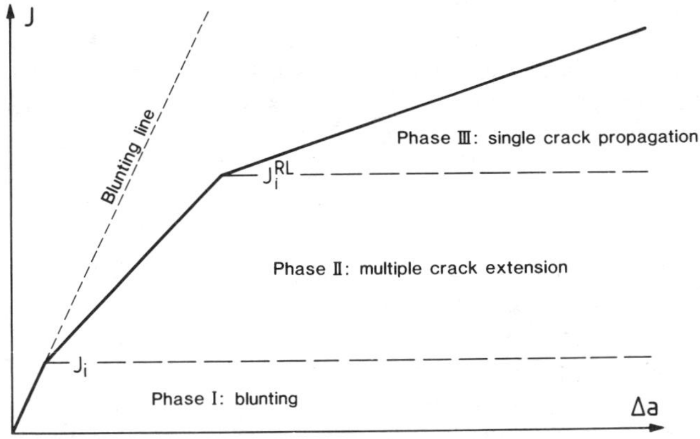

FIG. 10—*Modified* J-R *curve (schematic).*

The results presented here indicate that J_i^{RL} is lowest at low loading rates and for small specimens. Therefore, it seems to be reasonable to use statically loaded small specimens, for example, compact specimens, if lower-bound ductile fracture resistance curves are to be determined. The value of J_i^{RL} then will be so close to J_i that both kinks may no longer be distinguished, resulting in lower-bound *J-R* curves with a continuously changing slope.

On the other hand, J_i^{RL} increases with loading rate and especially with specimen size. Presumably, the kink in the resistance curve could be resolved for the first time so clearly in this investigation because two effects, each promoting Phase II and high values of J_i^{RL}, were combined. First, relatively large specimens were used, and second, the tests were performed at high loading rates. The high toughness level of this material, as indicated by the CT 25 data, may contribute to this effect, too. Also, the special specimen orientation (L-S) might be responsible for this effect.

Finally, note that most of the data of the impacted SENB 90 specimens exceed the J capacity, J_{max}. But side-crack formation seems to start at the end of blunting and, therefore, at J levels well within ASTM limits. Single-crack propagation, however, occurs only outside of these limits despite the large specimen size with the high J capacity of 1580 KN/m. Whereas the J capacity is exceeded by most of the data of the SENB 90 specimens, the ω criteria as proposed in Ref *15*,

$$\omega = \frac{W - a}{J} \frac{dJ}{da} \gg 1 \tag{4}$$

is fulfilled up to the pronounced kink in the J-resistance curve. Up to this kink, ω is decreasing continuously from $\omega > 100$ to $\omega \approx 15$ (Phase II), but thereafter ω is dropping to $\omega \approx 4$ (Phase III). Therefore, at Phase III, that is, for single-crack propagation and the flat part of the J-resistance curve, neither the ASTM validity requirements nor the ω criteria are fulfilled. On the other hand, at the beginning of Phase II, that is, for multiple crack extension and the steep part of the *J-R* curve, at first both criteria are fulfilled and at the end of Phase II at least the ω criteria are fulfilled.

In this context, it may be mentioned that, especially for the impacted SENB 90 specimens, the maximum of the measured load (initial peaks are not considered here) coincides with the kink of the *J-R* curve. Therefore, in these investigations, the fracture mode below plastic collapse is correlated with multiple crack formation, whereas the propagation of a single ductile crack occurs only under plastic collapse situation.

Of particular interest for the evaluation of components is the fact that the above-defined critical value for single-crack propagation, J_i^{RL}, increases with increasing specimen size. Therefore, Phase II with the corresponding steeper resistance curve could be important for larger components. If the trends reported here could be confirmed, additional safety margins may be expected for components.

Summary

The investigations aimed at the characterization of the ductile fracture of a pressure vessel steel at elevated loading rates are based on the *J*-integral concept. To this end, impact experiments were performed with SENB 90 specimens in a drop-weight tower and with Charpy specimens in a pendulum. All specimens were fatigued and side-grooved.

From the measured hammer load signals, *J*-integral values were determined. By a multispecimen technique, the available impact energy was varied to achieve different amounts of crack extension, and dynamic *J-R* curves were determined. The *J-R* curve of the SENB 90 specimens lies significantly above related static and dynamic results. Also, a significant change in slope was found at $\Delta a \approx 3$ mm.

Investigations of the fracture surfaces and sectionings perpendicular to the fracture surface revealed that the stretched zone is followed by a region where the ductile main crack may be accompanied by side cracks up to about 3 mm in length. Two-crack systems as well as three-crack systems have been observed. The length of the side cracks corresponds approximately to the crack length at which the *J-R* curve exhibits a change in slope. Examinations of other experiments indicate that the tendency to develop side cracks and, therefore, to increase the fracture resistance increases with the loading rate and, in particular, with the specimen size.

Acknowledgment

The author is grateful to the Bundesministerium für Forschung und Technologie (BMFT) and the Deutsche Forschungs-Gemeinschaft (DFG) who supported parts of these investigations.

References

[1] Angelino, G. and Fossati, C., "Dynamic Elastic-Plastic Characterization of Nuclear Pressure Vessel Material," in *Advances in Fracture Research*, Vol. 5, D. Francois, Ed., Pergamon Press, New York, 1981, pp. 2345–2352.

[2] Brüninghaus, K., Hesse, W., Twickler, M., Twickler, R., and Dahl, W., "The Influence of Loading Rate on *J-R* Curve of 20 MnMoNi 5 5 Steel using the Key Curve Method," *Journal de Physique*, Coll. C5, Suppl. 8, Tome 46, 1985, pp. 227–232.

[3] Curr, M. R., MacGillivray, H. J., and Turner, C. E., IMechE Seminar on Dynamic Fracture Mechanics, London, 1986.

[4] Kobayashi, T., Yamamoto, I., and Niiomi, M., *Engineering Fracture Mechanics*, Vol. 24, 1986, pp. 773–782.

[5] Joyce, J. A. and Hackett, E. M., "Application of the Key Curve and Multiple-Specimen Techniques to Dynamic *J-R* Curve Testing of Alloy Steel," NUREG/CR-4579, U.S. Nuclear Regulatory Commission, Washington, DC, 1986.

[6] Packeiser, R., Aurich, D., and Ziebs, J., "Influence of Loading Rate on the Crack Resistance Curve and the Initiation Temperature," *Nuclear Engineering and Design*, Vol. 102, 1987, pp. 487–493.

[7] Blumenauer, H. and Ortmann, R., "Determination of Dynamic Crack Initiation Toughness by the Instrumented Precracked Charpy Test," in *Proceedings of the European Conference on Fracture*, ECF 6, Amsterdam, 1988, pp. 441–450.

[8] Böhme, W., "Experimentelle Untersuchungen zum elastisch-plastischen Bruchverhalten eines Druckbehälterstahls unter dynamischer Beanspruchung," Scientific Report W 3/89, Fraunhofer-Institut für Werkstoffmechanik (IWM), Freiburg, Jan. 1989.

[9] Kalthoff, J. F., Böhme, W., Winkler, S., and Klemm, W., "Measurements of Dynamic Stress Intensity Factors in Impacted Bend Specimens," presented at Committee on the Safety of Nuclear Installations (CSNI) Specialists Meeting on Instrumented Precracked Charpy Testing, Electric Power Research Institute (EPRI), Palo Alto, CA, 1980, in EPRI NP-2102-LD, 1981, pp. 1–17.
[10] Böhme, W. and Kalthoff, J. F., "The Behavior of Notched Bend Specimens in Impact Testing," *International Journal of Fracture*, Vol. 20, 1982, pp. R139–R143.
[11] Böhme, W., "Experimentelle Untersuchungen dynamischer Effekte beim Kerbschlagbiegeversuch," Scientific Report W 1/85, IWM, Freiburg, 1985.
[12] Böhme, W., "Dynamic Key-Curves for Brittle Fracture Impact Tests and Establishment of a Transition Time," in *Fracture Mechanics: Twenty-First Symposium, ASTM STP 1074*, J. P. Gudas, J. A. Joyce, and E. M. Hackett, Eds., American Society for Testing and Materials, Philadelphia, 1990, pp. 144–156.
[13] Böhme, W. and Kalthoff, J. F., "On the Quantification of Dynamic Effects in Impact Loading and the Practical Application for K_{Id}-Determination," *Journal de Physique*, Coll. C5, Suppl. 8, Tome 46, 1985, pp. 213–218.
[14] Vo, Thi and Stöckl, H., "Simulationen von Schlagversuchen an großen Biegeproben aus einem Druckbehälterstahl mit der Methode der finiten Elemente," Scientific Report W 2/89, IWM, Freiburg, 1989.
[15] Hutchinson, J. W. and Paris, P. C., "Stability Analysis of *J*-Controlled Crack Growth," in *Elastic-Plastic Fracture, ASTM STP 668*, J. D. Landes et al., Eds., American Society for Testing and Materials, Philadelphia, 1979, pp. 37–64.

DISCUSSION

Mark T. Kirk[1] *(written discussion)*—During his presentation, Dr. Böhme empirically correlated the existence of "side-crack" development in elastic-plastic fracture toughness tests to increase in specimen size and specimen loading rate for the particular material tested. Increases in these two parameters indicate an elevated constraint, or stress triaxiality, condition at the crack tip. Any comment the author can make regarding theoretical reasons for *why* "side cracks" should form (that is, showing a link to constraint, shear band formation, logarithmic spiral formation, and so forth) would serve as helpful guidance to other researchers. Additionally, comments as to the effect of crack plane orientation on "side-crack" formation would be helpful.

Wolfgang Böhme (author's closure)—The dependence of side-crack formation on size and loading rate suggests, indeed, a correlation with the crack-tip constraint. To answer the question "why," additional experiments and theoretical as well as numerical investigations for a more detailed analysis of the crack-tip stress fields are planned or underway. I would consider further statements concerning the reason for side-crack formation too speculative.

T. Varga[2] *(written discussion)*—Böhme's findings on crack branching can be confirmed: Crack branching in quasi-static testing of CT specimens of ASTM A508 ?? steel (of French origin) up to 10-in. (250-mm approximately) wall thickness was observed as a rule. Therefore, high K_{Ic} values were measured and a more plastic behavior was observed than with a single crack.

In several cases in which a great number of small cracks were formed in an envelope (similar to a plastic fan), a pseudoplastic behavior was observed (designed and published).

Wolfgang Böhme (author's closure)—I am pleased to learn that obviously similar effects have also been observed in other investigations. However, I have not been able to find corresponding references, so I would be very grateful if those references could be identified by the discusser.

[1] David Taylor Research Center, U.S. Navy, DRTC, Code 2814, Annapolis, MD 21402.
[2] TVFA TU-Vienna, Karlsplatz, Austria A1040.

Paul McConnell[1]

Dynamic Fracture Toughness of Ductile Iron

REFERENCE: McConnell, P., **"Dynamic Fracture Toughness of Ductile Iron,"** *Rapid Load Fracture Testing, ASTM STP 1130,* Ravinder Chona and William R. Corwin, Eds., American Society for Testing and Materials, Philadelphia, 1991, pp. 104–117.

ABSTRACT: Dynamic *J*-integral fracture-toughness test results were obtained for several different ductile irons over a range of temperatures. The ductile irons were all of a chemistry and microstructure in nominal compliance with ASTM Specification for Ferritic Ductile Iron Suitable for Low Temperature Service (A 874).

Among the materials investigated were ductile irons from actual nuclear spent-fuel storage casks and from a cask to be used in a drop-test program at a national laboratory. These ductile irons were of a mainly ferritic matrix with highly spherical graphite nodules. The tests were performed over a range of temperatures in an attempt to define the dynamic fracture-toughness transition temperature regime, important because the service temperature for transport casks for "normal" conditions can be as low as −40°C.

The technique employed for measuring the dynamic *J*-integral value was based on the multiple-specimen technique described in ASTM Test for J_{Ic}, a Measure of Fracture Toughness (E 813). Details of the test technique are discussed.

The upper-shelf dynamic fracture-toughness values for the three ductile irons investigated were similar, ranging from 61 to 77 MPa · m$^{1/2}$ at room temperature.

The transition between upper-shelf and lower-shelf dynamic fracture toughness occurred over the range of approximately −35 to −50°C for one ductile iron and the range of −50 to −68°C for another. These transition temperature regimes were higher than the quasi-static transition temperature noted for similar ductile iron material.

The influence of microstructure on the fracture toughness of ductile iron is discussed.

KEYWORDS: fracture toughness, dynamic, ductile (cast) iron, ferritic, nodular (cast) iron, spent-fuel casks

With the enactment of the Nuclear Waste Policy Act in 1982 [*1*], the U.S. federal government became responsible for the disposition of spent nuclear fuel from commercial nuclear power plants. The federal government will take possession of this spent fuel at the power plant site and transport the material to its ultimate, permanent destination in a licensed geological repository (presently scheduled to open in 2010). Temporary spent-fuel storage is an additional consideration: some utilities may require additional on-site storage facilities if their existing pool storage capacity becomes exhausted before the repository is in operation.

One option for storage of spent fuel is the use of large "casks" to contain intact or consolidated fuel assemblies. These casks may be placed alongside the reactor or may be transported to any away-from-reactor site for temporary or permanent dry storage. Indeed, spent fuel is presently being stored in the United States at nuclear utility sites by this means. The dual-purpose cask function of both storage and transportation offers some logistical and technical advantages.

Manufacturers are designing a new generation of thick-walled monolithic spent-fuel casks that might serve the dual purpose of transport and storage. These casks, for economic reasons, may be

[1] Formerly, General Research Corp., Santa Barbara, CA 93111; presently, GRAM, Inc., 8500 Menual Blvd. N.E., B370, Albuquerque, NM 87112.

made of ferritic materials. Ferritic materials can, under certain conditions, fail by means of a brittle fracture process. (Transport casks licensed to date generally have been constructed of relatively expensive austenitic stainless steel, which fails only in a ductile manner.)

Among other conditions, the U.S. Code of Federal Regulations [*2*] requires, as a prerequisite to licensing a cask for spent-fuel transport application, that it be demonstrated that the cask be able to withstand a 9-m drop onto an essentially unyielding surface at an orientation that would inflict maximum damage. Another stipulation is that there be no loss of cask integrity after a 1-m drop onto a 0.15-m-diameter steel pin (puncture test), again in a position for which maximum damage is expected. Both the impact and puncture events represent severe challenges to ferritic materials, especially because the temperature at which such "hypothetical accidents" are assumed to occur is −29°C.

An attractive ferritic material for use in spent-fuel casks is ferritic ductile iron. All cast irons are ferrous alloys containing carbon, silicon, magnesium, and low levels of sulfur and phosphorus. Excess carbon precipitates as free graphite of a quantity normally between 8 and 12 volume/%. The graphite can take various shapes including flakes or spheroids (nodules) in a matrix consisting of some combination of ferrite, pearlite, austenite, or martensite. Of interest is "ductile" iron with an essentially pure ferrite-matrix and highly spherical graphite nodules due to its relatively high toughness and ductility.

Because of the possibility of brittle failure of ductile iron, or any ferritic material, a discipline of importance in the design of spent-fuel casks is fracture mechanics. Fracture mechanics links three key variables that control brittle fracture: namely, applied stress, flaw size, and the fracture toughness of the component material. The combination of flaw size and applied stress result in a stress intensity that can, if it exceeds the material fracture toughness, result in crack initiation, crack growth, and perhaps unstable brittle crack propagation.

The fracture toughness of a ferritic material, such as ductile iron, can decrease with decreasing temperature or increasing loading rate. At a given loading rate, the fracture toughness decreases from a relatively high "upper-shelf" value at some elevated temperature through a transition range to a "lower-shelf" value at lower temperatures symptomatic of a brittle fracture failure mode. The transition range can be narrow, as has been demonstrated for ductile iron at quasi-static loading rates [*3*], or quite broad, as for certain steels.

Research has been sponsored recently to quantify the fracture toughness of ferritic ductile irons [*3–7*]. Motivation for research on ferritic ductile iron was, in part, due to the paucity of fracture-toughness data for the material, its relatively low fracture toughness in comparison with better characterized ferritic steels, and the potential economic advantage of the material.

For ferritic ductile irons with moderate levels of silicon content, the quasi-static loading rate fracture-toughness transition from upper to lower shelf occurs at temperatures below −70°C [*3*]. In addition, evidence suggests that the quasi-static upper-shelf fracture toughness of ferritic ductile iron is directly related to the graphite nodule shape and spacing [*8*].

Increasing the loading rate may result in a shift in the transition temperature to a higher temperature. Therefore, it is important to have information on the fracture toughness of ductile iron at the appropriate temperature and loading rate because casks could, should an accident occur, be subject to high loading rates at low temperatures.

Test Program

An objective of this program, sponsored by Sandia National Laboratories [*5,6*], was to measure the high loading rate fracture toughness/temperature relationship for ferritic ductile irons with a range of microstructures possibly suitable for use in spent-fuel transport casks. An additional objective was to relate the measured fracture-toughness values to the microstructure of the

TABLE 1—*Ductile iron mechanical, chemical, microstructural properties.*[a]

Ductile Iron	YS, MPa	UTS, MPa	Elongation, %	K_{Jc} at −29°C, MPa · m$^{1/2}$	Ferrite,[b] %	Nodule Count, mm^{-2}	Nodule Spacing, μm	C, w/o	Si, w/o	S, w/o	Ni, w/o
A	247	367	29	76	88.2	147.5	41	3.54	2.16	0.003	0.51
P1[c]	238	379	27	95	86.4	97.1	51	3.56	1.72	0.006	0.06
P3[c]	249	372	21	95	83.4	37.7	82	3.39	1.74	0.005	0.05
P5[c]	250	377	16	98	78.5	24.1	102	3.32	1.70	0.005	0.06
E	N/A	N/A	N/A	100	87.0	21.5	108	3.41	1.93	0.004	0.25

[a] All materials meet ASTM Specification A 874. Tensile and quasi-static fracture-toughness values from Ref *16*. Fracture-toughness values converted from *J*-integral values.

[b] Pearlite contents less than 5%; balance is graphite.

[c] Tensile values at −30°C, others at 20°C.

ductile irons and to compare the microstructure/property relationships with those identified in the previously referenced ductile iron database [*8*].

The ductile irons evaluated were supplied by a spent-fuel cask vendor. The ductile irons were taken from actual spent-fuel casks. The materials are listed in Table 1 with some of their mechanical, microstructural, and chemical properties. All of the ductile irons were mainly ferritic, possessed good nodularity, and were of high quality in terms of their quasi-static fracture toughness. As such, the ductile irons met the requirements of ASTM Specification for Ferritic Ductile Iron Castings Suitable for Low Temperature Service (A 874). This specification limits the silicon level to no more than 2.3 w/o to ensure a relatively low transition temperature.

To assess the effect of microstructure, and in particular the graphite nodule spacing, on the fracture toughness of ductile iron at high loading rates, it was desirable to test ductile irons with the widest differences in nodule spacing. Nine candidate ductile irons were available in sufficient quantity for a test program. Those listed in Table 1 met the criteria of high ferrite content, spherical graphite nodules, and a range of nodule spacing.

The ductile irons, identified P1, P3, and P5, were from a spent-fuel cask (MOSAIK KfK) to be used for drop test demonstrations at Sandia National Laboratories. Material was taken from the 21.6-cm-thick bottom of the cask at three through-wall locations; P1 was from the inner wall, P3 from the mid-wall, and P5 from the outer wall.

High Loading Rate Fracture Toughness Tests

ASTM Test for J_{Ic}, a Measure of Fracture Toughness (E 813) three-point bend bar fracture-toughness specimens 12.7 mm thick and 25 mm wide were machined from the ductile irons selected for high loading rate fracture-toughness testing. The specimens were fatigue precracked to a crack length/width ratio of approximately 0.55. The specimen size was dictated by the available material. The thickness chosen was estimated to be sufficient, however, to provide size-valid J-integral test results.

Multiple specimen J-integral tests were performed over a range of temperatures to define the fracture toughness/temperature relationship. The test procedure for the impact J-integral tests was based on that specified in ASTM Test E 813. There were certain necessary differences from the technique described in ASTM Test E 813, however. Most notable was the fact that the tests were performed at a dynamic rate where the tests occurred within a few milliseconds; E 813 specifies quasi-static loading rates of from 0.1 to 10.0 min. Drop tests of casks have demonstrated that peak loading occurs within 10 to 40 ms, so the loading rate used was conservative.

The load, P, was applied to the test specimens with a drop tower system equipped with a dynamic load cell tup. Semiconductor strain gages are embedded within the face of the striking tup. General procedures for instrumented impact testing may be reviewed in Refs *9* and *10*.

In the procedure used, a series of specimens was loaded to different deflections, at a given temperature, resulting in different amounts of stable crack growth. The deflections were controlled in the high loading rate fracture-toughness test system by placing "stop blocks" alongside and slightly below the test specimen (Fig. 1). When the impacting instrumented tup struck the stop blocks, deflection of the specimen and stable crack extension within the specimen were interrupted (Fig. 2). A brake system eliminated rebounding of the drop weight back onto the specimen. By placing the blocks at slightly different positions below the top of each specimen of a series of bend bars, different deflections and crack extensions were achieved.

The displacement, δ, of the bend bars was measured with a linear variable displacement transducer (LVDT) attached to the tup and was recorded on the digital recorder. The LVDT used was a Schaevitz Model PCA-116-300 with a nominal linear range of 0.6 in. (15 mm). The LVDT was calibrated to an accuracy of approximately 0.001 in. (0.254 mm) using a micrometer system attached to the crosshead of the drop tower.

FIG. 1—*Drop tower apparatus for dynamic* J*-integral testing showing test specimen, anvils, stop blocks, instrumented tup and LVDT, before impacting specimen.*

The load output from the tup and displacement from the LVDT were recorded on a high-speed LeCroy digital recorder. The sampling rate for tests on ductile irons A and E was 30 kHz, but for the P1, P3, and P5 tests it was some 50 times faster; the load and displacement data were sampled at 1 MHz. The faster sampling rate provided a more detailed record of the impact event.

The load and displacement signals were subsequently cross-plotted to provide the load versus load-line displacement curve for each J-integral test. The output of the LVDT was filtered, resulting in a delay as compared with the load signal of approximately 325 μs. To eliminate the associated error, this delay was compensated for in the cross-plotting of the load and displacement signals by shifting the displacement signal to the appropriate time.

As the result of oscillations inherent to the impact load record, the point at which the tup struck the stop blocks was occasionally difficult to define. However, because the allowable displacement of the test specimens was physically measured before the test, this point could be identified on the load-displacement plot by comparing the time for an equivalent displacement from the LVDT record with the corresponding time on the load record.

Trapezoidal numerical integration was performed on each load-displacement curve. Integration of this curve is necessary for the calculation of the J-integral, which is proportional to the area under the P-δ curve (the energy). The J-integral value for each test was calculated per the expression provided in ASTM Test E 813 based on the energy absorbed by the test specimen. The apparent energy under the P-δ curve may include some contribution due to the compliance of the test machine and inertial loading associated with the impacting tup. This contribution is believed to be very small, however, due to the stiffness of the test machine and the relatively low energy required to deflect the specimens compared to that imparted to the specimens.

The crack extension, Δa, and the original fatigue precrack length for each specimen were

FIG. 2—*Drop tower apparatus after striking specimen.*

measured post-test. The specimens were heat-tinted and fractured after they were immersed in liquid nitrogen. A nine-point average measure of the crack extension was made. Little crack tunneling was noted. Some crack extension may have occurred in the test specimens after the striking tup hit the stop blocks. This would result in an overestimate of specimen crack extension for a given value of the J-integral. This is discussed further in the section entitled "Test Results."

The J-integral values, J_I, for a series of specimens at a given temperature, were subsequently plotted against the corresponding crack extensions, Δa, thereby generating the J-integral crack growth resistance curve (J-R curve) for the ductile iron at that temperature. From the J-R curve, the initiation value of the J-integral, J_{Id}, was derived by extrapolating a linear regression fit to the individual $J_I/\Delta a$ datum points back to the crack-tip blunting line of the ductile irons. A value of 275 MPa was assumed for the flow strength, σ_0, of the ductile irons for the purpose of plotting the blunting line. (This flow strength is representative of that for ferritic ductile iron at quasi-static loading rates at nominal ambient temperatures; higher values have been noted at higher rates/lower temperatures [*4,8*]. The differences have a small effect on the blunting line slope, the intersection with the R-curve, and J_{Id} value, however.) A power-law fit was also made to the J-Δa pairs and the J_{Id} derived from the intersection of this curve with the 0.2-mm offset line. Because of the difficulty in controlling the amount of crack extension in a given test specimen by means of the subtle adjustment of the deflection allowed by the stop blocks, the region for valid data and the spacing between J-Δa data pairs specified by ASTM Test E 813 was often relaxed.

The J_{Id} values have been converted to an estimate of the fracture toughness, K_{Jd}. This conversion, $K_{Jd} = (J_{Id} \cdot E)^{1/2}$, used a conservative value of 160 000 MPa for the ductile iron elastic modulus, E.

The stress intensity rate was estimated by dividing the K_{Jd} value by the loading time associated with the test specimen with a J_I value most nearly equal to the derived J_{Id} value.

TABLE 2—*Ductile iron high loading rate fracture-toughness results.*

Ductile Iron	Temperature, °C	J_{Id} Linear-Regression, kJ/m²	Power-Law, kJ/m²	J_{Id}, Adjusted for Inertia, kJ/m²	K_{Jd}, MPa · m$^{1/2}$	K_Q,[a] MPa · m$^{1/2}$
A	18	24	. . .		61	
	−29	24	39		61/79	
	−40	21	15		57/49	
	−46	16	17		50/51	
	−51					39
	−57					42
	−62					46
	−73					48
	−196					39
E	18	35	43		74/83	
	−29	21			58	
	−40	21	19		58/55	
	−46	21	18		58/54	
	−51					46
	−62					40
	−73					29
	−196					25
P1	18	10	44		39/83	
	−29	2	20		17/57	
P3	18	38	46	52	55/85/91	
	−29	17	23	37	51/61/77	
P5	18	32	47		72/87	
	−29	21	33		58/73	

[a] K_Q values do not meet ASTM Test E 399 size requirements for 12.7-mm-thick specimen. J_{Id} values *do* meet ASTM Test E 813 size requirements for 12.7-mm-thick specimen.

At lower temperatures, unstable crack initiation occurred. This behavior precluded construction of the *J-R* curve. Instead, a linear-elastic-based fracture-toughness value, K_Q, was calculated for these specimens based on the point at which a rapid drop in load, indicating "pop-in," occurred. This calculation was in accordance with the expression provided in ASTM Test for Plane-Strain Fracture Toughness of Metallic Materials (E 399).

The test specimens were subjected to chemical and microstructural analyses post-test for correlation of the fracture-toughness test results with microstructural properties.

Test Results

High Loading Rate Fracture-Toughness Data

Table 2 shows the high loading rate fracture-toughness test results on ductile irons A and E and the ductile iron from Plates 1, 3, and 5. A typical load-versus-deflection plot is shown in Fig. 3. This figure was obtained by cross-plotting the load versus time and LVDT deflection versus time signals. For each ductile iron, there is a set of such curves for each test temperature. Each test of a given set at a given temperature represents a slightly different amount of deflection and crack extension. The *J*-integral associated with the point at which the striking tup hit the stop blocks was calculated from each of the load-deflection curves. The plastic and elastic components of the *J*-integral were calculated for ductile irons A and E; the total energy was used in the calculation of the *J*-integral for ductile irons P1, P3, and P5.

For each test series, J_I-Δa was plotted and the data fit with a linear-regression curve and a

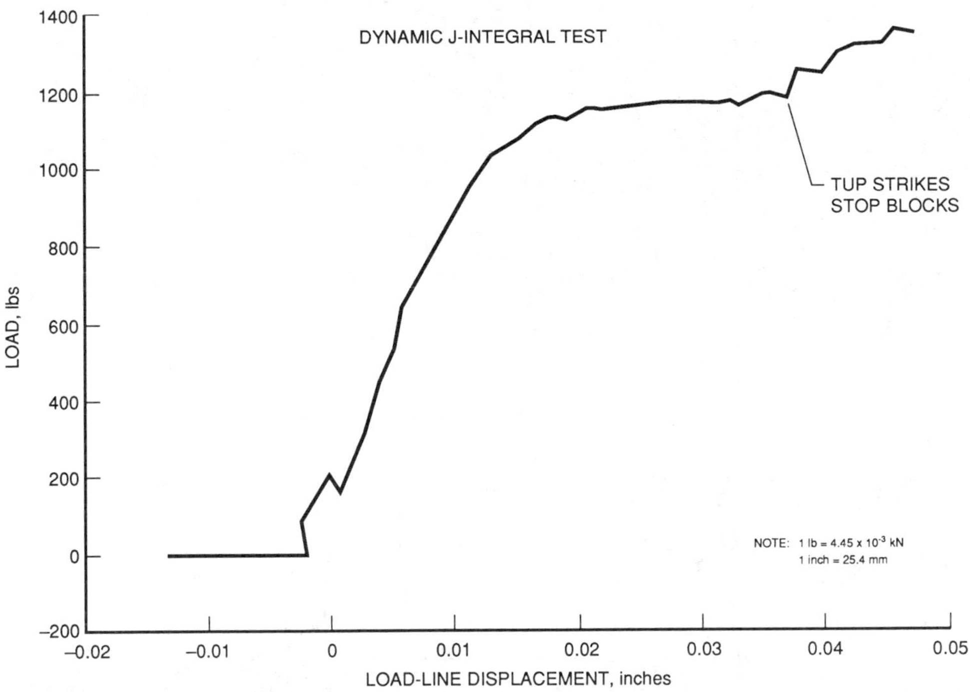

FIG. 3—*Typical load-versus-displacement curve for high loading rate* J-*integral test.*

power-law curve. Figures 4 and 5 are representative J-integral versus crack growth resistance curves for ductile iron A at a temperature of −29°C (linear-regression and power-law fit, respectively). Some scatter is evident in the J-Δa pairs. This may be due to the material variability between specimens tested at a given temperature or to test technique. From these curves, the crack initiation value of the J-integral was determined. This value represents the energy required to induce stable crack growth under high loading rate conditions. The values for J_{Id} for all of the ductile irons are tabulated for each temperature in Table 2. All of the J_{Id} values reported in Table 2 meet the ASTM Test E 813 size validation requirements (B, $b_0 > 25\ J_Q/\sigma_Y$). The J_{Id} values have been converted to an equivalent linear-elastic parameter, K_{Jd}. The value derived for J_{Id} for a particular test from the J-Δa curve fits depended on whether the linear-regression or power-law fit to the data was used. Occasionally, values for J_{Id} obtained from the linear-regression or power-law fit were unrealistically low in that an actual J_I-Δa datum point would exist near the blunting line. This point corresponded to a specimen for which there was only minor crack extension, a specimen exhibiting near crack initiation conditions. From the associated absorbed energy by such a specimen a more realistic value of the J_{Id} was used for calculation of the fracture toughness, K_{Jd}, and is listed in Table 2.

The J_{Id} values have been converted to an equivalent linear-elastic parameter, K_{Jd}, so that the upper-shelf data could be compared with the lower-shelf K_Q values to show the fracture toughness versus temperature transition of the ductile irons at high loading rates. Note that ASTM Test E 813 cautions against the application of this conversion, particularly when small specimens are used for J integral testing since a change in fracture mode may occur with larger specimens.

These fracture toughness versus temperature curves provide a measure of the transition tem-

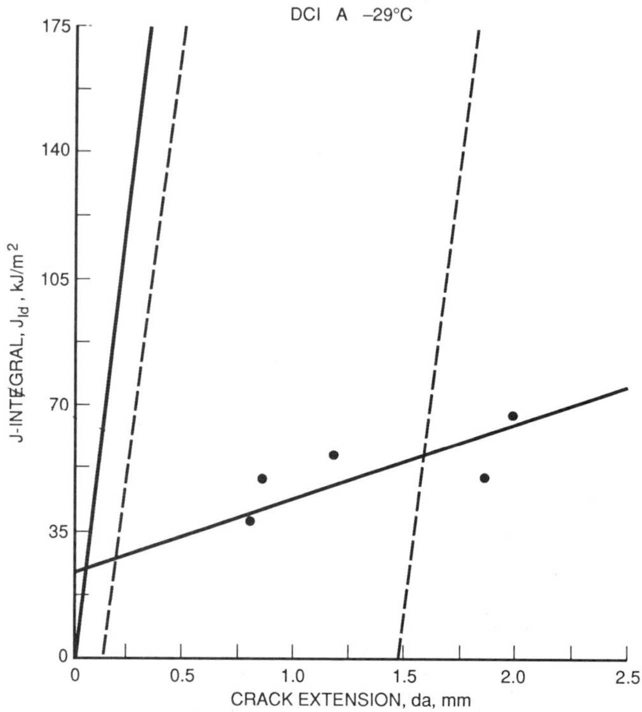

FIG. 4—*Linear-regression fit to* J *versus* Δa *data for ductile iron A at −29°C.*

perature at high loading rates and the fracture toughness of these ductile irons at the intended service temperatures (for example, −29°C). The high loading rate fracture-toughness values (K_{Jd}) for ductile irons A and E at −30°C were 61 and 58 MPa · $m^{1/2}$, respectively, at a stress intensity rate of approximately 10^5 MPa · $m^{1/2}$/s. For ductile iron A, the transition from upper shelf to lower shelf occurred over the range of approximately −35 to −51°C. For ductile iron E, this range was approximately −50 to −70°C.

The lower-shelf fracture-toughness value was determined by testing specimens of each material at −196°C. The resultant K_Q value was 39 MPa · $m^{1/2}$ for ductile iron A and 25 MPa · $m^{1/2}$ for material E. These K_Q values do not meet the size criteria for linear-elastic fracture toughness, the test specimens being only 12.7 mm thick, and they may not be equivalent to K_{Id}, the true lower-shelf dynamic fracture toughness of the materials.

One potential source of inaccuracy in the test results may be the inertia of the test specimen after the instrumented tup has struck the stop blocks. The measure of the energy used to calculate the value of the *J*-integral associated with the point at which the tup strikes the stop blocks may not coincide with the actual crack extension of the test specimen, incurred post-test, should the specimen continue to deflect even after leaving contact with the tup. This effect is schematically illustrated for ductile iron P3 in Fig. 6. Assuming inertia of the test specimen results in extra deflection and crack extension to the point of maximum load applied to the impacting striker, the extra energy associated with the actual measure of crack extension can be estimated from the load-deflection curve. This results in a higher value of J_I for a given crack extension (Fig. 7) and, therefore, a higher value of the estimated crack initiation parameter, J_{Id}. The magnitude of this potential inaccuracy is reported in Table 2 for ductile iron P3 at 18 and −29°C.

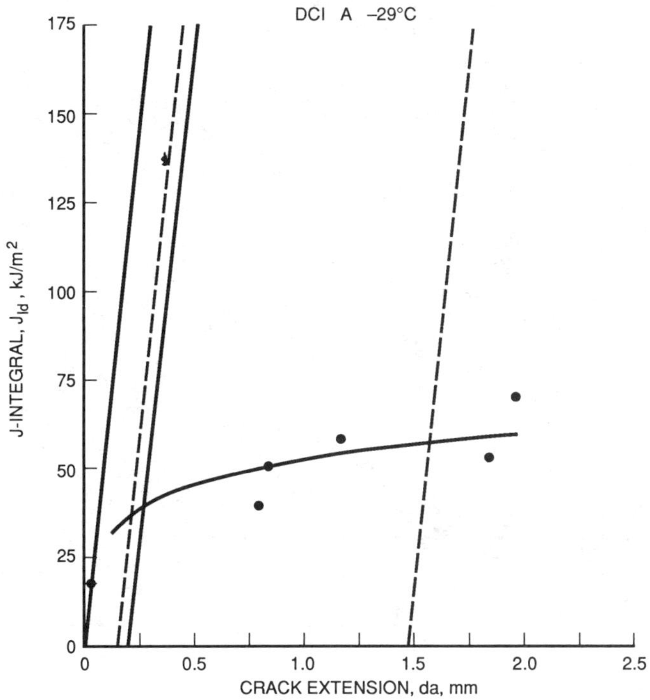

FIG. 5—*Power-law fit to* J *versus* Δa *data for ductile iron A at −29°C.*

In a separate study by Joyce[2] on a different but very similar ferritic ductile iron conforming to ASTM Specification A 874, dynamic fracture toughness values (K_{Jd}) averaging 107 MPa · m$^{1/2}$ were obtained at 23°C and 96 MPa · m$^{1/2}$ at −40°C. The test technique used [*11*] differs from the one described herein. Joyce used a double crack-opening displacement gage to identify the onset of crack initiation. The specimen was completely fractured during a single-specimen *J*-integral test. Specimen displacement was measured optically. The relatively low values of fracture toughness determined for the ductile irons in the present study may be due to the use of a different test technique.

A previous study on ductile iron A [*12*] indicates that the quasi-static fracture toughness was upper shelf and ductile crack initiation was observed at temperatures as low as −85°C. The K_{Jc} value at −30°C was 76 MPa · m$^{1/2}$. Even at an intermediate loading rate ($\simeq 10^3$ MPa · m$^{1/2}$/s), the −30°C fracture toughness was 90 MPa · m$^{1/2}$, which was presumed to be upper shelf because the crack initiation was ductile and the magnitude of the fracture-toughness value approximated the quasi-static value. These values are compared to a −30°C fracture toughness of 61 MPa · m$^{1/2}$ at the loading rate used in the present study (10^5 MPa · m$^{1/2}$/s). This behavior, whereby the magnitude of the upper-shelf fracture toughness at the very high loading rates may be somewhat lower than at reduced loading rates, requires a mechanistic explanation. The same behavior has been observed for ductile iron E in that the quasi-static fracture toughness (K_{Jc}) at −30°C (upper shelf) was determined to be 101 MPa · in$^{1/2}$ and the intermediate loading rate (10^3 MPa · in$^{1/2}$/s)

[2] J. A. Joyce, Mechanical Engineering Department, U. S. Naval Academy, private communication, May 1989.

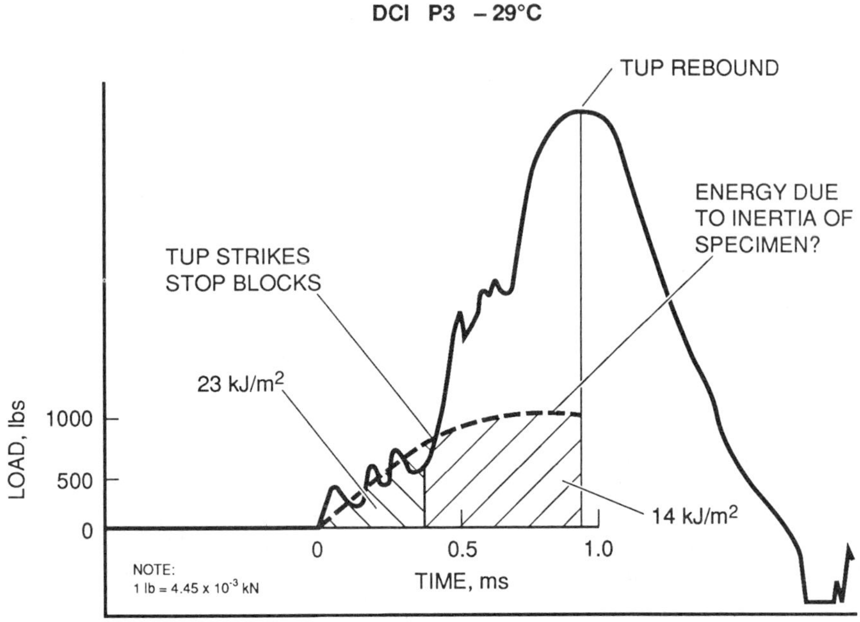

FIG. 6—*Representation of loading event for J-integral test showing possible extra energy absorbed due to specimen inertia.*

value was 110 MPa · $m^{1/2}$ at −30°C. In the present study, the K_{Jd} at −30°C was determined to be 58 MPa · $m^{1/2}$, again less than at the lower loading rates.

It is often assumed that ferritic steel materials have a higher upper-shelf fracture toughness at dynamic loading rates. Joyce and Hackett [*13*] have noted that for ferritic A106 steel that the dynamic J_I values are not elevated over the quasi-static rate value. They suggest this observation may be due to the tendency for the higher loading rate to lower the ductility of the A106 steel and to cause crack initiation and extension with less bend angle. It may be the case, too, that the micromechanisms for ductile fracture in ferritic ductile iron, described in detail for quasi-static loading rates by Voigt and Eldoky [*14*], are rate-dependent in that, above some threshold, loading rate ductility and crack initiation resistance are diminished, resulting in a decrease in the high loading rate fracture-toughness value. Voigt and Eldoky define five events that lead to ductile macrocrack initiation in ferritic ductile iron:graphite nodule decohesion, localized plastic deformation in the matrix, matrix microcracking, microcrack link-up, and link-up of micro-cracks with macrocrack front. The process of plastic deformation in the matrix region between graphite nodules may be particularly rate-sensitive, thereby affecting the dynamic fracture toughness of the material.

A test program conducted at Battelle Columbus Laboratories has provided crack-arrest fracture-toughness data (K_{Ia}) on the same ductile iron herein identified as ductile iron A [*15*]. Crack-arrest fracture toughness is believed to represent a lower-bound fracture-toughness value at a given temperature, and extremely high loading rate fracture-toughness values, such as generated in this study, should approach the K_{Ia} values. Unlike the *J*-integral values in the present study, the majority of the Battelle K_{Ia} data were invalid due to insufficient specimen size. However, a valid K_{Ia} datum point was obtained at −40°C, which can be compared with the K_{Jd} value obtained in the present study. The valid K_{Ia} value was approximately 55 MPa · $m^{1/2}$, which

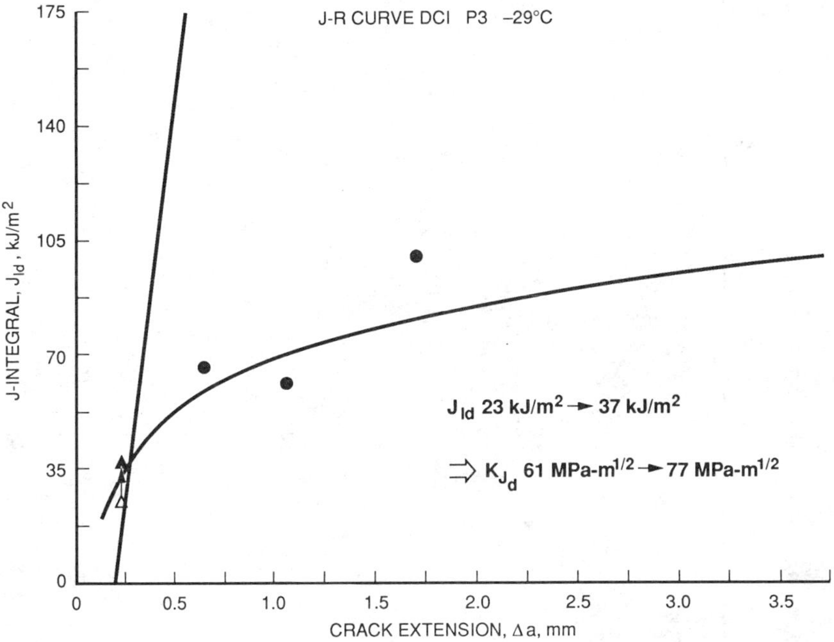

FIG. 7—J-R *curve for ductile iron P3 at −29°C showing correction for specimen inertia (note triangular data points).*

is very close to the K_{Jd} values generated on ductile irons A and E. The reported K_{Ia} values increased at temperatures above −40°C, however, exceeding the K_{Jd} values obtained in the present study. This anomaly may be due to the fact that the 25-mm-thick K_{Ia} specimens were not thick enough to provide size-valid results. However, investigation of the possible differences between K_{Ia} and K_{Jd} at the upper-shelf temperatures is warranted.

Microstructure/Fracture Toughness Relationship

A linear relationship between upper-shelf quasi-static fracture toughness and graphite nodule spacing has been identified for ferritic ductile iron [*3*]. A similar relationship may exist between the upper-shelf dynamic fracture toughness and graphite nodule spacing, but is not obvious over the narrow range of nodule spacings investigated. The present K_{Jd} (linear-regression) data are plotted on Fig. 8 with data from Ref *8* for comparison.

Ductile iron E, with the higher fracture toughness, had an average nodule spacing of approximately 110 μm. Ductile iron A had a narrower spacing of approximately 40 μm and a lower fracture toughness. This difference in the upper-shelf fracture toughness among all of the ductile irons of the range of spacing investigated did not exhibit a marked trend, however.

There is also evidence that the shape of the graphite nodules in ductile iron affects the fracture toughness. In general, spheroidal nodules are thought to promote higher fracture toughness. Elongated, linear nodules apparently facilitate crack initiation and subsequent growth [*14*]. ASTM Specification A 874 for ferritic ductile iron requires a minimum of 90% spheroidal nodules to enhance fracture toughness. The ASTM Specification for Evaluating Microstructure of Graphite in Iron Castings (A 247) designates six types of graphite "flakes." Spheroidal nodules

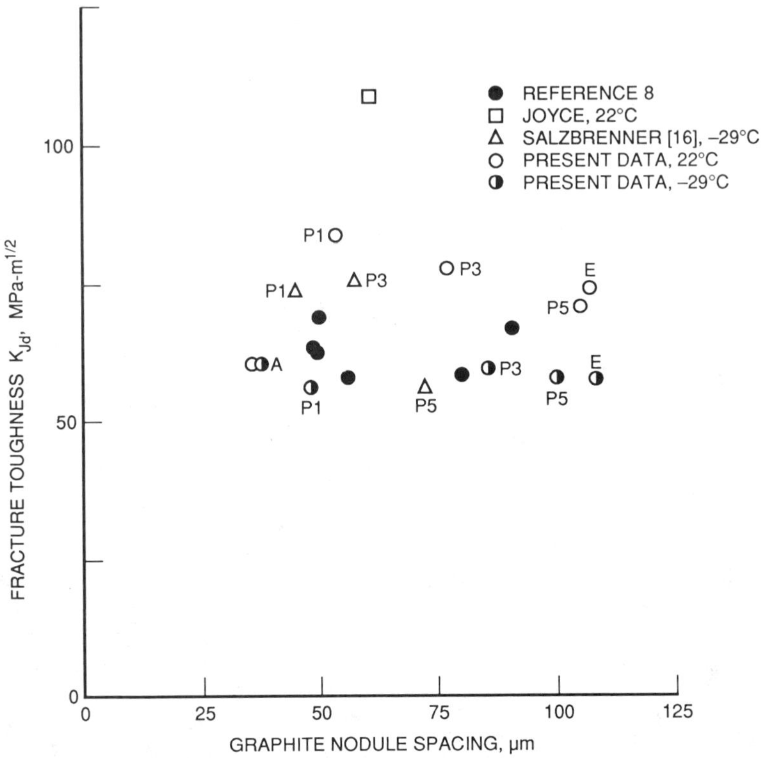

FIG. 8—*Dynamic fracture-toughness values versus ductile iron graphite nodule spacing.*

are denoted as Types I and II. Type VI nodules are essentially exploded spheroids and are not elongated. Apparently, Type VI nodules are not significantly detrimental to fracture toughness because it has been observed that other ductile irons of relatively high fracture toughness also have some Type VI nodules and less than 90% Types I and II nodules [*8*]. Some of the P-series ductile iron also had less than 90% Types I and II nodules. The ductile iron from Plate 5 had a significant percentage (to 40%) of Type VI nodules, with the remainder being mostly Types I and II. Plate 3 material had up to 30% Type VI nodules. However, the fracture toughness of the ductile iron from Plate 5 was essentially equivalent to that from Plates 1 and 3.

Conclusions

Dynamic fracture-toughness testing (stress intensity rate of approximately 10^5 MPa · m$^{1/2}$/s) was performed on ductile irons over a temperature range of +20 to −196°C to define the magnitude of the ductile upper-shelf fracture toughness and the transition temperature regime. These ductile irons were ferritic and with good graphite nodularity, typical of the microstructure meeting ASTM Specification A 874. The testing closely followed the requirements of ASTM Test E 813 for *J*-integral fracture-toughness testing.

The upper-shelf dynamic fracture-toughness values for the three ductile irons investigated were similar, ranging from 61 to 77 MPa · m$^{1/2}$ at room temperature. The transition between upper- and lower-shelf dynamic fracture toughness occurred over the range of approximately −35 to −50°C for one ductile iron (A) and the range of −50 to −68°C for another (ductile iron E). These transition temperature regimes were higher than the quasi-static transition temperature noted for similar ductile iron material [*3*]. The difference in fracture toughness between two of the ferritic ductile irons may have been due to the difference in graphite nodule spacing.

Although all of the irons had highly nodular graphite, ductile irons P3 and P5 had considerable ASTM Type VI nodules with no apparent detriment to the fracture toughness of the material.

The test results must be evaluated with caution in that the fracture-toughness values reported may not be the true initiation parameters. The data may be subject to the vagaries of the experimental test technique for measuring high loading rate *J*-integral fracture toughness. The relatively small size of the test specimens (12.7 mm thick), although large enough to meet *J*-integral size requirements, warrants consideration when extrapolating data to large components. The data are, nevertheless, representative of some measure of the fracture toughness of the materials investigated. The apparent low upper-shelf fracture toughness of the ductile irons at the elevated loading rate as compared with the quasi-static fracture-toughness values remains to be explained.

References

[*1*] The Nuclear Waste Policy Act of 1982, Public Law 97-425.

[*2*] Office of Federal Register, *Title 10, Code of Federal Regulations,* Parts 71 and 72, Washington, Jan. 1984.

[*3*] Salzbrenner, R., "Fracture Toughness of Ferritic Ductile Cast Iron," Report SAND86-05635, Sandia National Laboratories, Albuquerque, 1986.

[*4*] McConnell, P. and Sheckherd, J. W., "High Loading Rate Fracture Toughness of Nodular Ductile Cast Iron," Report SAND85-7259, Sandia National Laboratories, Albuquerque, Dec. 1984.

[*5*] McConnell, P. et al., "An Investigation of the Dynamic Fracture Toughness Transition Temperature of Ferritic Ductile Iron," Report TTC-0824, Sandia National Laboratories, Albuquerque, Dec. 1988.

[*6*] McConnell, P. et al., "Upper-Shelf Dynamic Fracture Toughness of Ferritic Ductile Iron from a Mosaik Cask," Report TTC-0928, Sandia National Laboratories, Albuquerque, Sept. 1988.

[*7*] Rosenfeld, A. R. et al., "Crack Arrest Toughness of Nodular Iron," submitted to *Journal of Nuclear Engineering and Design,* 1989.

[*8*] McConnell, P. and Lombrozo, P., "Ductile Iron Data Base—Correlations Between Microstructure and Fracture Toughness," Report SAND86-7263, Sandia National Laboratories, Albuquerque, Feb. 1987.

[*9*] DeSisto, T. S., Ed., *Instrumented Impact Testing, ASTM STP 563,* American Society for Testing and Materials, Philadelphia, 1974.

[*10*] Wullaert, R. A., Ed., "C.S.N.I. Specialist Meeting on Instrumented Precracked Charpy Testing, Proceedings," Report NP-2107-LD, Electric Power Research Institute, Nov. 1981.

[*11*] Joyce, J. A. and Hackett, E. M., "Dynamic *J-R* Curve Testing of a High Strength Steel Using the Key Curve and Multispecimen Techniques," in *Fracture Mechanics, Seventeenth Volume, ASTM STP 905,* American Society for Testing and Materials, Philadelphia, 1986, p. 741.

[*12*] McConnell, P. and Sheckherd, J. W., "Fracture Toughness Characterization of a Ductile Iron Spent Nuclear Fuel Cask," Report for RP2717-02, Electric Power Research Institute, Palo Alto, CA, June 1986.

[*13*] Joyce, J. A. and Hackett, E. M., "Transition Range Drop Tower *J-R* Curve Testing of A106 Steel," NRC NUREG/CR-4818, Feb. 1987.

[*14*] Voigt, R. C. and Eldoky, L., "Microstructural Aspects of Fracture in Ductile Cast Irons," in *Proceedings,* Advanced Casting Technology International Conference, American Society for Metals, Metals Park, OH, 1987.

[*15*] Rosenfield, A. R. et al., "Crack Arrest Toughness of Nodular Iron," Battelle Memorial Institute, Columbus, Report for RP-2813-12, Electric Power Research Institute, Palo Alto, CA, April 1988.

[*16*] Salzbrenner, R., "Mechanical Property Mapping of the Ductile Cast Iron MOSAIK KfK Cask," Report SAND90-0776 Sandia National Laboratories, Albuquerque, Aug. 1990.

Robert L. Tregoning,[1] *Jason M. Shapiro,*[2] *and William N. Sharpe, Jr.*[1]

Dynamic Crack-Tip Opening Displacement (CTOD) Measurements with Application to Fracture Toughness Testing

REFERENCE: Tregoning, R. L., Shapiro, J. M., and Sharpe, W. N., Jr., **"Dynamic Crack-Tip Opening Displacement (CTOD) Measurements with Application to Fracture Toughness Testing,"** *Rapid Load Fracture Testing, ASTM STP 1130,* Ravinder Chona and William R. Corwin, Eds., American Society for Testing and Materials, Philadelphia, 1992, pp. 118–133.

ABSTRACT: Crack-tip opening displacement (CTOD) measurements were made 100 μm behind the crack tip on standard three-point bend specimens using a laser-based interferometric technique. Four materials (SAE-01 steel, 4340 steel, HY-100 steel, and tungsten) were tested at loading rates ranging from quasi-static ($\dot{K} \approx 1$ MPa $\cdot$ m$^{1/2}$/s) to dynamic ($\dot{K} \approx 10^6$ MPa $\cdot$ m$^{1/2}$/s).

The quasi-static and intermediate rate tests and analyses were conducted according to the guidelines specified in ASTM Test for Plane-Strain Fracture Toughness of Metallic Materials (E 399). In these lower rate tests, load was applied by a servohydraulic testing machine and a clip gage was used to record the crack-mouth opening displacement (CMOD). From these data, the stress intensity factor, K, and the fracture toughness can easily be determined from the standard methodology.

In the high rate tests, inertial effects are prevalent and K is not as easily measured. The initial measurements of CTOD, however, can be used to provide a calibration between the static and dynamic tests if K is assumed to increase linearly with time during the 10 to 15 μs before the crack starts to propagate. If this assumption holds, dynamic time-CTOD results can be converted to K-CTOD curves using a companion specimen technique. Further, the dynamic fracture toughness can then be found by superimposing static and dynamic K-CTOD results and employing a reduced slope technique to locate the onset of crack propagation.

This dynamic technique has been modeled after existing practices defined in ASTM Test E 399 and leads to experimental scatter comparable to that found in static fracture-toughness tests. The results show that there is no dramatic change in the fracture toughness over this range of loading rates for any of the materials studied.

KEYWORDS: crack-tip opening displacement, dynamic fracture, fracture toughness, interferometry, steels, impact

As the rate of loading applied to a precracked fracture-toughness specimen increases, the interpretation of the measured load and crack-mouth displacement becomes increasingly difficult. Inertial, or wave propagation, effects come into play and mean that the relation between the far-field applied load and the near-field parameter (for example, the stress intensity factor K) is no longer simple or precise. A further complication to rapid-load testing is the difficulty in determining the critical event when the crack starts to propagate. The ideal test would simultaneously measure stress and crack position at the crack tip and take less than 1 ms to conduct.

[1] Graduate research assistant and professor, respectively, Department of Mechanical Engineering, The Johns Hopkins University, Latrobe Hall 127, Baltimore, MD 21218.

[2] Patent advisor, U.S. Army Laboratory Command, Adelphi, MD. Formerly, Department of Mechanical Engineering, The Johns Hopkins University, Latrobe Hall 127, Baltimore, MD 21218.

Such measurements are either not possible (in the case of stress) or are very difficult (in the case of position). However, crack-tip opening displacements (CTOD) can be measured rather easily on simple specimens, and that is the theme of this paper.

Laser-based interferometry between two reflecting indentations placed across the crack and within 100 μm of its tip was used to measure the CTOD of three-point bend (3PB) specimens. Static tests were conducted in an electrohydraulic test machine with the applied load and crack-mouth opening displacements (CMOD) measured simultaneously with the CTOD. Dynamic tests, in which the specimen was impacted by a projectile, measured only the dynamic CTOD; the duration of these tests was approximately 10 μs. One would expect a large variation among the CTOD tests because precracks in specimens cannot be reproduced exactly and the indentations are only 100 μm behind the tip of a crack in a specimen that is 6350 μm thick. However, it was found that the variation among CTOD measured on different specimens subjected to identical loadings was acceptable.

Conversion of the measured CTOD to a useful measure of resistance to fracture requires nontrivial assumptions about the dynamic load application and the rate-dependence of the material. A procedure is presented here that uses these assumptions and leads to the determination of a dynamic fracture toughness. The values seem reasonable and consistent with values measured by other techniques. Our position is that the dynamic CTOD measurements are accurate and can possibly be used to make a more definitive determination of dynamic fracture toughness. However, it may be necessary to combine these measurements with auxiliary techniques such as a three-dimensional finite-element analysis that accounts for material rate effects to improve the precision of the results.

Much of this paper is taken from an extensive report [*1*] that describes the experiments in detail, but the first author conducted additional experiments that agreed closely with the earlier work. After a brief review of other high-rate techniques, the basics of the laser-based interferometry system are presented. Four materials (SAE-01 steel, 4340 steel, HY-100 steel, and tungsten) were tested, and the static and dynamic test procedures are described. Not only are the techniques for measuring CTOD given, but the procedures for computing fracture toughness are outlined. The results are presented in the form of plots of static and dynamic K versus CTOD for the four materials. The closing remarks discuss the potential and the drawbacks of CTOD measurements for rapid load fracture testing.

Background

Conducting material tests at high loading rates presents many problems to the experimentalist. In a dynamic environment, the effects of wave propagation and inertia must be accounted for in the design and evaluation of a test method. Furthermore, the phenomena of crack initiation is a very localized event, and measurements taken at locations other than the crack tip must be interpreted with a great deal of care. Finally, laboratory instrumentation imposes limits on the rate, reliability, and cost of a given test method. For these reasons, a variety of experimental techniques have been studied over the years with the intention of establishing a standard procedure for dynamic fracture-toughness testing.

The standard method for evaluating fracture toughness under quasi-static loading conditions, ASTM Test for Plane-Strain Fracture Toughness of Metallic Materials (E 399), involves a precracked specimen of simple geometry and simultaneous recording of the applied load and CMOD. The test is generally conducted in a servohydraulic testing machine. The dynamic tests employ similar specimens for the most part, but vary widely in their methods of loading and the types of measurements made. High-rate hydraulic testing machines, compressed gas guns, Charpy impact testing machines, and explosives have all been used to induce the rapid loading conditions necessary in the study of dynamic fracture. Usually an incident load pulse is recorded

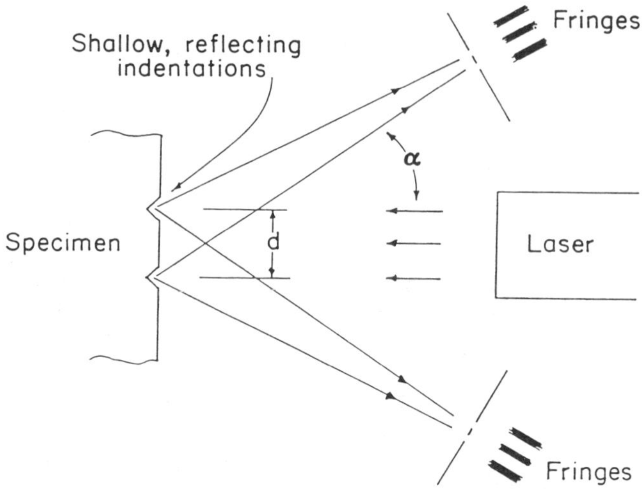

FIG. 1—*Schematic of the interferometric strain-displacement gage.*

and then some measurement is made of a physical quantity related to the onset of crack initiation. This could be CMOD, strain near the crack tip, surface dimpling ahead of the crack, or CTOD.

Rapid-load fracture tests in a hydraulic test machine are similar to quasi-static tests [*2*]. ASTM Test E 399 has even been extended to allow specimens to be tested using conventional quasi-static measurement techniques as long as the loading pulse rise time is longer than 1 ms. The disadvantages of this extension, however, are the relatively low loading rates that can be generated ($\dot{K} \approx 10^4$ MPa · m$^{1/2}$/s) and the fact that displacement records are not as smooth in the linear range as those obtained in static tests.

In higher rate tests, many researchers have chosen to instrument their specimens with strain gages located near the crack tip [*3–6*]. Strains recorded dynamically can then be used to construct stress intensity factor histories either by assuming the strain distribution to be the same as in the quasi-static case or by performing a dynamic finite-element analysis for a given specimen geometry and gage location [*6*]. An advantage to this method is its relative simplicity; it is possible to use any of the aforementioned loading techniques to obtain rates up to 10^6 MPa · m$^{1/2}$/s. It is necessary, however, to place the strain gage a distance away from the plastic zone immediately ahead of the crack tip.

Many novel optical methods have been used successfully in the study of dynamic fracture as well. One unique method involves a long rod with an annular precrack [*7,8*]. The specimen is loaded explosively to obtain rates greater than 10^6 MPa · m$^{1/2}$/s. The rod is instrumented with strain gages and an optical gage for measuring crack-opening displacement. A major drawback to this technique is the size of the specimen (25 mm in diameter by 1 m long) and the need for special precracking equipment. Other optical methods include caustics [*9–11*], Moiré interferometry [*12*], speckle photography [*13,14*], and high-speed photography of microgrids [*15*]. All of these techniques require fairly elaborate and expensive instrumentation.

CTOD Measurement Technique

The interferometric strain-displacement gage (ISDG) is a laser-based technique for measuring the relative displacement between two reflecting indentations on a specimen surface. Figure 1 is

TABLE 1—*Material properties for all materials tested.*

Material	Elastic Modulus, 1000 MPa	0.2% Yield Stress, MPa	Hardness, Rockwell C
SAE-01	213	345	14
4340	226	1530	52
HY-100	223	765	24
Tungsten	316	1130	39

a schematic illustrating the basic principle. The indentations are impressed into the specimen a small distance apart with a Vickers microhardness tester. When an indentation is illuminated by a coherent and monochromatic light source normal to the specimen surface, light is diffracted from each of the four triangular sides. The Fraunhofer diffraction patterns emanating from the two closely placed indentations interfere to form fringes in space. This is simply an example of Young's two-slit interference phenomenon in reflection rather than transmission. For constructive interference to occur, the difference in optical path must equal an integer multiple of the wavelength of the incident light. This can be written as

$$d \sin \alpha = m\lambda \tag{1}$$

where λ is the wavelength of the laser, α is the angle between the incident laser beam and the observation position, d is the distance between the centers of the indentations, and m is the fringe order. The relation between the fringe motion and the relative displacement of the two indentations is given by

$$\delta d = \frac{\delta M_1 + \delta M_2}{2} \frac{\lambda}{\sin \alpha_0} \tag{2}$$

where α_0 is the angle between the incident laser beam and the fixed observation points and δM_1, δM_2 are the relative fringe motions of the fringe patterns in line with the direction of indent displacement. "Relative fringe motion" means the displacement of a fringe divided by the distance between it and the next fringe. If one places a detector at a fixed position, α_0, and observes the intensity to change from dark to bright, then one has observed a change in fringe order of one half. This corresponds to a relative surface displacement of ~½ μm. A more extensive discussion of the optical principles is given in [*16*].

Materials and Specimens

Four materials were tested in this study: SAE-01 tool steel, 4340 steel, HY-100 steel, and tungsten. The tool steel was used in the development of the test methods. The dynamic experiments in particular required several tests to perfect the triggering and recording instrumentation and procedures. More data of a similar nature are available for 4340 steel than any other material, so it was tested to provide a comparison with the results of other researchers. The tungsten and HY-100 stock were provided by Aberdeen Proving Ground and David Taylor Research Center, respectively. Table 1 shows the material properties for all the materials tested.

Standard three-point bend specimens were used in both the quasi-static and dynamic fracture-toughness tests. Figure 2 gives the geometry and dimensions of the specimens tested. In each of these specimens, the starter notch is oriented parallel to the rolling direction and the specimen

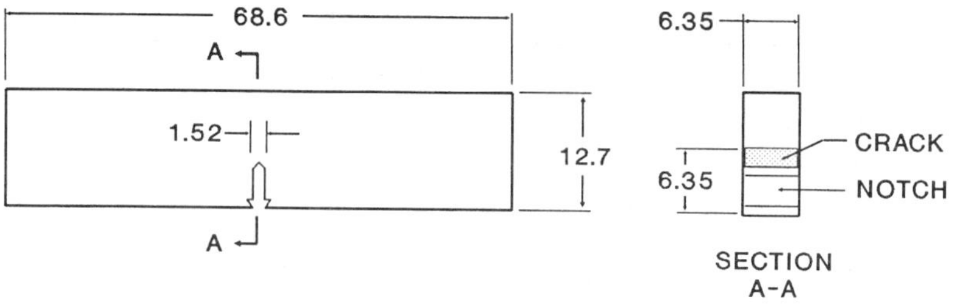

FIG. 2—*Dimensions of the three-point-bend specimens—50 mm anvil spacing in all cases.*

width is parallel to the long transverse direction of the plate stock material. This is the T-L starter notch orientation as designated by ASTM. The precracking procedures followed the ASTM Test E 399 guidelines closely, keeping the applied stress intensity levels below 60% of the anticipated critical value. The precracking is important because, as will be seen, the closure load is a reasonable fraction of the applied load, and uniformity of the closure loads is preferred.

After the specimens were machined and precracked, they were polished using various grits of sandpaper and finally a 0.5-μm diamond paste. Indents were then applied with a LECO microhardness tester at a standardized position 100 μm behind the crack tip and 70 μm apart. The two angles for each fringe pattern were measured by placing each specimen within a protractor-screen arrangement and illuminating it with a small laser. The angles were measured from a line normal to the surface of the specimen to the center of the fringe pattern.

Static Experimental Procedures

The frequency response of the measurement devices is not a concern in the quasi-static tests. Measurements can therefore be made using standard techniques to determine load, CTOD, and CMOD. The equipment and acquisition techniques used in these tests as well as the method for determining the static fracture toughness from the raw data are described in this section.

The quasi-static measurement system (Fig. 3) consists of an electrohydraulic testing machine, a microcomputer system for data acquisition and test control, and the sensors required to measure the load, the CTOD, and the CMOD. The test is begun when a digitized linear loading ramp is sent from the microcomputer to the actuator of the testing machine via the microcomputer's digital/analog (D/A) converter. As the three-point bend specimen is loaded, a standard load cell measures the load sensed by the tup and a clip gage records the CMOD. Simultaneously, two photomultiplier tubes (PMT) record the movement of the interference patterns emanating from the set of indentations located 100 μm behind the crack tip. The PMT signals, along with the clip gage and load cell readings, are sent to the microcomputer through appropriate analog-to-digital (A/D) channels for storage and further processing. Loading is continued until specimen failure occurs, at which point the experiment is terminated.

The clip gage, load cell, testing machine, support fixtures, and specimen all comply with ASTM Test E 399. The testing procedures and subsequent analysis also follow the guidelines set by this procedure. In essence, these experiments are merely those of classical fracture-toughness testing along with an additional measurement of the crack-tip opening displacement.

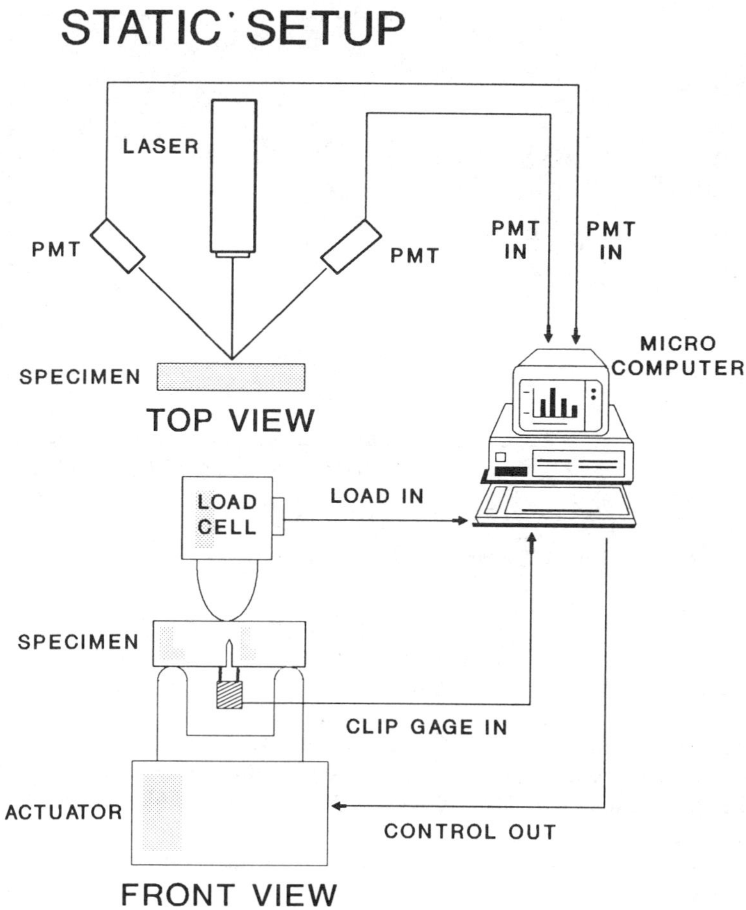

FIG. 3—*Schematic of the setup for the static tests.*

Once the tests are completed, the maximum and minimum peaks of the interference signals are found and converted into CTOD using Eq 2. These CTOD are next plotted as a function of the tup load, P, for each test. Then, individual PMT signals are averaged to cancel the effects of rigid body motion in the direction parallel to the long axis of the specimen. Typical raw data of load versus CTOD are shown in Fig. 4 for two static tests of tungsten. Note, in that figure, that P does not start at zero because of the closure effect. This load usually differs somewhat for each specimen depending on the history of the fatigue precracking.

Because each of the specimens tested generally has a different closure load, it is important to shift the CTOD results to a common origin for comparison. This transformation is achieved using the following method. First, a linear least-squares fit is applied to the first four points of CTOD data (corresponding to an opening displacement of ≈ 2 μm). This region represents the initial linear elastic response of the crack tip, and the least-squares technique yields an average elastic slope. The P-versus-CTOD curve is then extrapolated back to the abscissa (CTOD axis) using this elastic slope. Next, each data point is shifted along the CTOD axis until the intersection of the extrapolated curve and the CTOD axis (zero load) rests at the origin of the P-CTOD

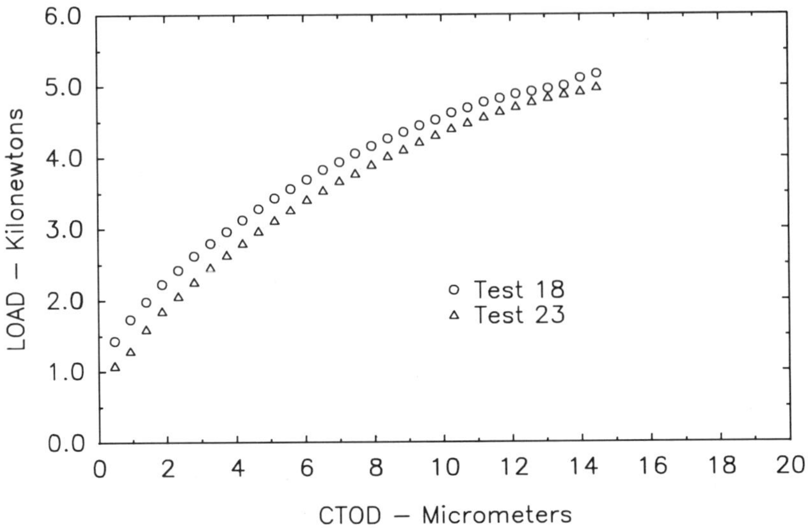

FIG. 4—*Raw load versus CTOD data for two static tests on tungsten.*

plot. Finally, the load values are rescaled in terms of the stress intensity factor by multiplication using the factor K/P calculated for the average crack length in accordance with ASTM Method E 399 procedures. These steps produce a curve for each test relating K and CTOD.

Similar plots are generated for the CMOD as a function of the tup load. The elastic slope of each of these curves is measured graphically, and a line is drawn from the origin with a slope reduced by 5% of the elastic value. The point of intersection between this "reduced" line and the original load-CMOD curve is defined as P_q, the load at crack initiation. Once P_q has been determined, K_q, the stress intensity factor at this load, is calculated from the appropriate equation in ASTM Test E 399.

Dynamic Experimental Procedures

The quasi-static rate tests ($\dot{K} \approx 1$ MPa $\cdot$ m$^{1/2}$/s) were conducted and analyzed by standardized techniques. A further extension into the dynamic region ($\dot{K} \approx 10^6$ MPa $\cdot$ m$^{1/2}$/s) is impossible for two reasons. First, hydraulic testing machines cannot achieve loading rates that lead to a loading rate this large. Second, measurements of load by the load cell and crack-tip opening from a clip gage at the crack mouth are simply invalid and bear no easily definable relation to the near-tip stress and displacement fields. At loading rates this large, tup and specimen inertial effects are prevalent and it is essential to make measurements as close to the crack tip as possible so that wave propagation effects can be neglected. Because of these problems, the experimental methods and analysis employed for these high-rate tests have to be drastically altered so that meaningful results can be obtained. These methods and the subsequent data analysis are described in this section; Fig. 5 is a schematic of the test setup.

Much faster loading pulse rise-times are achieved by firing a 2.5-cm-diameter, 12-cm-long aluminum projectile from a gas gun at speeds up to 50 m/s. This projectile strikes a small steel tup (see Fig. 5), which then transmits the loading pulse to the three-point bend specimen. After impact, the broken specimen, projectile, and tup land in a foam-filled catchbox. Before impact, however, the projectile crosses two laser beams placed a fixed distance apart. Interruption of the

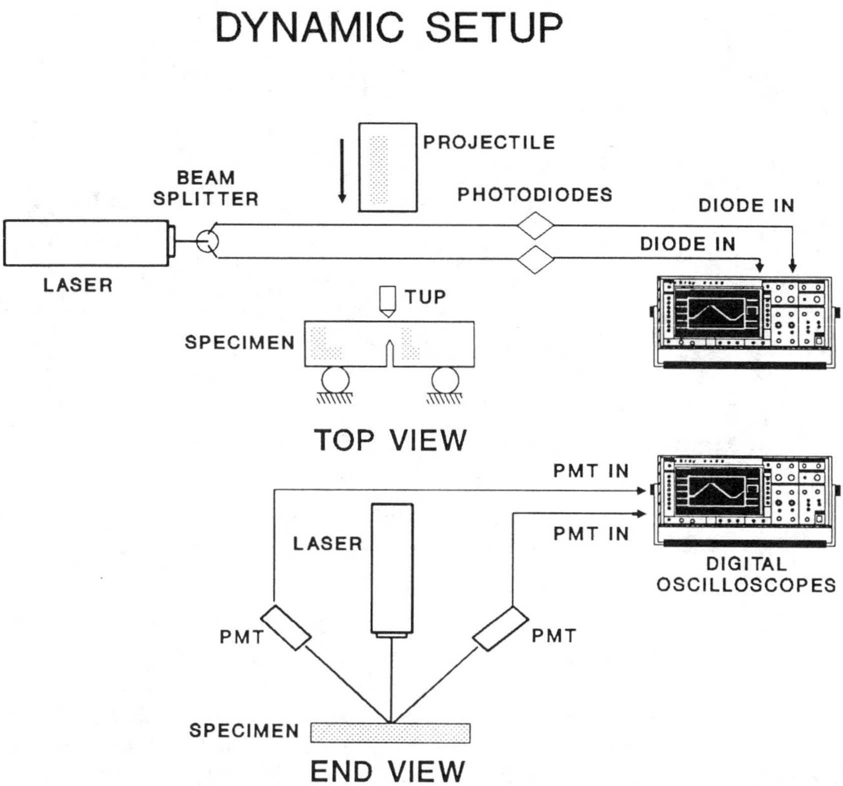

FIG. 5—*Schematic of the setup for the dynamic tests.*

light beam into the first photodiode triggers the start of data acquisition on a digital oscilloscope. Once the second beam has also been blocked, a permanent record of the output from both photodiodes has been captured. This record is analyzed after the test to determine the projectile's travel time over the distance spanned by the laser beams and, therefore, its velocity at impact.

A second laser illuminates the indentations on the specimen surface. These indentations are located 100 μm behind the crack tip. The fringe patterns from the specimen impinge on two RCA 8645 photomultiplier tubes (see Fig. 5), which are connected directly to a second digital oscilloscope. As the distance between the indentations begins to change, the change in output voltage from one of the PMTs is usually sufficient to trigger the data acquisition of the second scope. If not, later reflections of the incident laser beam from the crack's plastic wake occur approximately 260 μs after the initial fringe motion and are always large enough to serve as a trigger. The oscilloscope's memory spans 320 μs, and with proper positioning of the trigger delay, either of the events mentioned can start the oscilloscope's sweep and capture the fringe motion that occurs over approximately 10 μs. The PMTs have rise-times on the order of 2 ns, and the digital oscilloscope is set to collect data at its maximum rate of 10 ns per point. Therefore, the measurement system allows excellent resolution of the dynamic event.

After the test, the information from the PMTs is transferred to a microcomputer which analyzes the fringe signals in the same manner as for the static experiments. As in the static tests, the maximum and minimum voltage peaks are found and converted into a measure of the CTOD

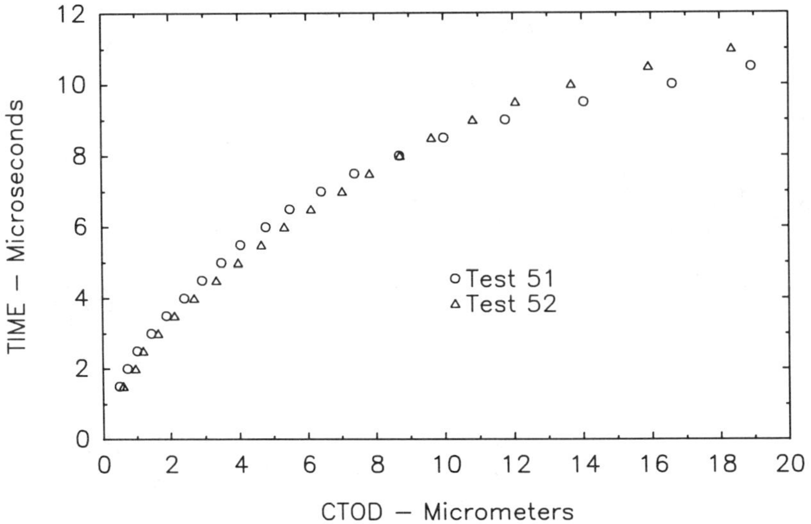

FIG. 6—*Raw time versus CTOD data for two dynamic tests on tungsten.*

using the method described earlier. However, these CTODs cannot be directly represented as a function of the load or stress at the crack tip as they were in the static case. This is because there is no reliable means to measure the stress at this position. Time is measured directly though, so it seems logical to present the raw data in this form. Figure 6 shows the dynamic CTOD results for two tungsten tests. Time, in this figure, is measured relative to the point of initial deviation of the fringe signal from horizontal, so the origin is different for each test.

The most encouraging aspect of these data is that each test is remarkably similar in appearance in both the initial linear elastic region and the more prominent plastic portion of the curve. Much more variation would generally be expected for measurements this close to the crack tip, considering the potential differences in the precracking history of each specimen. So, it seems that a viable and easily applicable experimental technique has been found to record the CTOD accurately in these high-rate tests.

It would be advantageous to interpret these CTOD results in a form that allows direct comparison with the complementary static CTOD curves for each material. Then, each material's fracture-toughness rate sensitivity could be studied. More importantly, a direct comparison between the results could ultimately lead to the formation of a material design parameter suitable for the determination of crack initiation in a dynamically loaded component. Obviously, for such a comparison to occur, the dynamic "time" variable must somehow be converted into the static K variable. The assumptions and method employed in this work to create such a transformation are now described.

First, the original data are shifted to the CTOD axis ($t = 0$) at the first measured CTOD point for $\Delta m = 1/2$. This step merely sets a common reference for each of the tests. Next, a least-squares linear fit is performed on the first four data points. This region corresponds to the linear response of the CTOD as determined visibly and yields a dynamic calibration factor. The crack-opening component of the stress field is assumed to rise linearly throughout the elastic and plastic portions of the loading until crack initiation. Also, the elastic response of the crack tip is assumed to be independent of the stress rate. If these assumptions hold, the time coordinate of the dynamic plots can be scaled by the ratio of the initial static CTOD slope computed earlier to

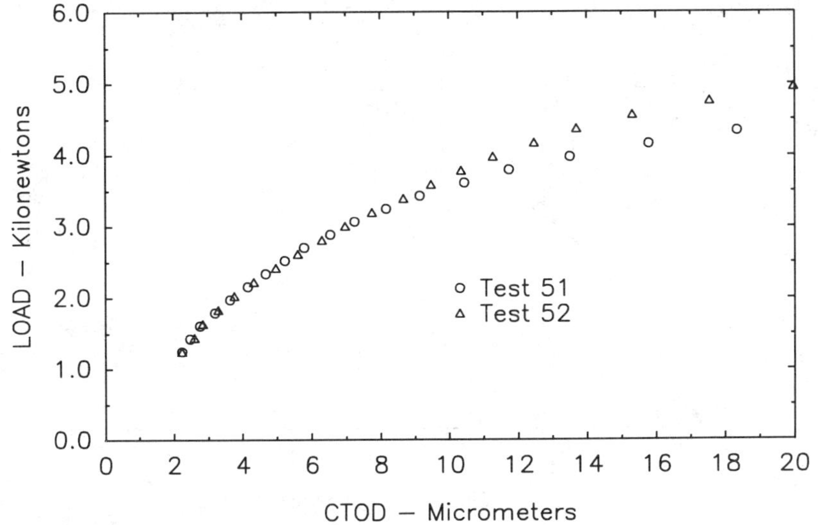

FIG. 7—*Final dynamic load versus CTOD data for the two tungsten tests.*

the dynamic calibration factor found above. Recall that the static slope has units of lb/μm whereas the dynamic slope has units of μs/μm. When combined, their ratio possesses units of lb/μs, which is then used to rescale the time axis to units of load. Of course, load cannot really be measured at the crack tip, and this intermediate relationship really relates the dynamic CTOD values to an equivalent static load that would be found at the tup if the test were somehow stopped at each measure of the CTOD and the stress field were allowed to stabilize.

The closure load must next be accounted for. This is easily measured before the dynamic test by loading each specimen statically until the PMT record an average fringe shift of 1/2. The load at this CTOD is taken as the closure load as in the static case. Because the crack-tip response is still elastic at this load, it can safely be regarded as the dynamic closure load. Each curve is shifted vertically by this amount.

Next, as in the static tests, the dynamic *P*-CTOD curve is extrapolated back to the abscissa (CTOD axis) using the slope of the elastic dynamic *P*-CTOD region. Each data point is then shifted along the CTOD axis until the intersection of the extrapolated curve and the CTOD axis rests at the origin of the dynamic *P*-CTOD plot. Again, this procedure sets a common origin for each of the dynamic plots. The final plots of *P* and CTOD for the two tungsten tests of Fig. 6 are shown in Fig. 7. Finally, the dynamic *P*-CTOD curves can be rescaled in terms of the stress intensity factor by multiplying the dynamic load by for each test by the appropriate ratio *K*/*P*; the results corresponding to the two tests in Fig. 7 are plotted in Fig. 8.

At this point, it may be helpful to summarize the process of obtaining dynamic *K*-versus-CTOD curves by the companion specimen technique.

1. Measure static load versus CTOD on a specimen. Figure 4 illustrates typical data. Use the linear elastic response to shift these data to a common origin.
2. Measure dynamic time versus CTOD for a specimen that is nearly identical to its companion specimen in the previous step. Figure 6 shows typical results.
3. Match the linear region of the *P*-CTOD curves (Fig. 4) with the linear region of the time-CTOD curves (Fig. 6) to develop a *P*-versus-*t* relationship.

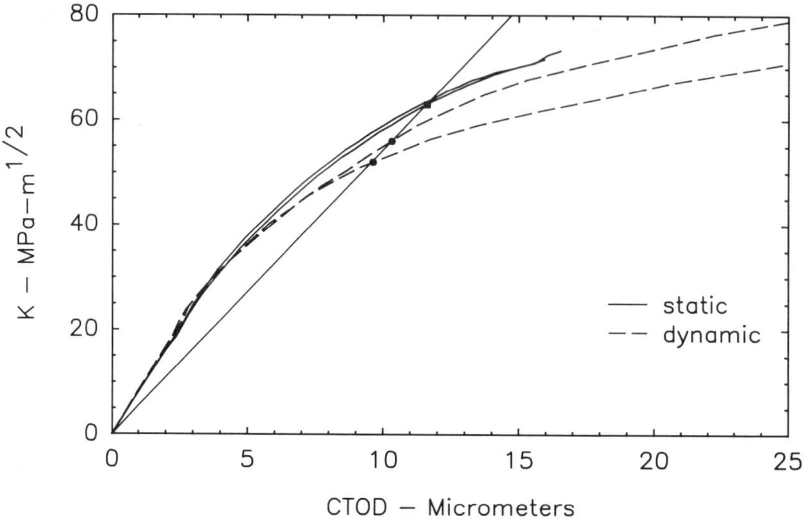

FIG. 8—*Static and dynamic* K *versus CTOD results for tungsten.*

4. Rescale the time axis to load using the above relationship. The first data point is fixed at the closure load for a dynamic specimen. This step is used to transform Fig. 6 into Fig. 7.
5. Finally, convert P to K to obtain Fig. 8.

In the results reported here, two or more specimens were tested to generate the linear P-t relationships from Steps 1 and 2. Therefore, this relation is an average for all specimens tested of the same material type.

Dynamic Fracture Toughness—Analysis and Results

The dynamic fracture toughness can be found using an analysis based on a combination of the static and dynamic CTOD results. In slower rate tests, ASTM specifies a reduced slope method to determine the crack initiation load, P_q, from the clip gage results. Therefore, if accurate clip gage measurements could be made for dynamic tests, it would be a simple matter to combine static, intermediate, and dynamic results on one graph, draw a single reduced slope, and determine a unique P_q for each test.

However, only CTOD measurements can be made accurately in the highest loading rate tests, but a reduced slope technique similar in form to the one prescribed by ASTM can be followed if one criterion is met. The crack lengths and the locations of the indentations for the static and dynamic tests must be nearly identical so that the initial elastic compliances will be the same. The dynamic K_q can then be determined in the following manner, assuming that the reduced slope method can be employed. The limitations involved with this assumption are presented later.

First, the static and dynamic K-versus-CTOD curves are plotted on the same graph. K_q for the static tests (which was obtained from the CMOD results) is plotted on the static CTOD curve, and a line is drawn from the origin to the static K_q values, which passes through the dynamic curves. The point of intersection between this line and the dynamic load-CTOD results gives the dynamic K_q.

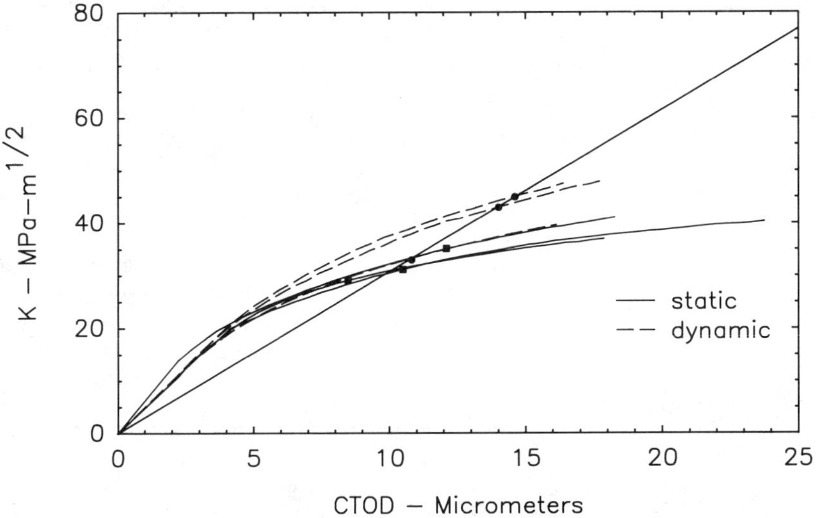

FIG. 9—*Static and dynamic* K *versus CTOD results for SAE-01 steel.*

The combined plots for the four materials are presented in Figs. 8 through 11. The K_q values are plotted as filled squares for the various static tests; these symbols show the local CTOD at which crack initiation (as defined by ASTM Test E 399) begins. A straight line is drawn through them using a visual fit; variability in the measured static K_q contributes to the uncertainty of the measurements. The intersection of this line with the dynamic curves is shown as a filled circle, and this corresponds to the dynamic K_q.

The three static CTOD plots of SAE-01 steel are close together, but the variability in establish-

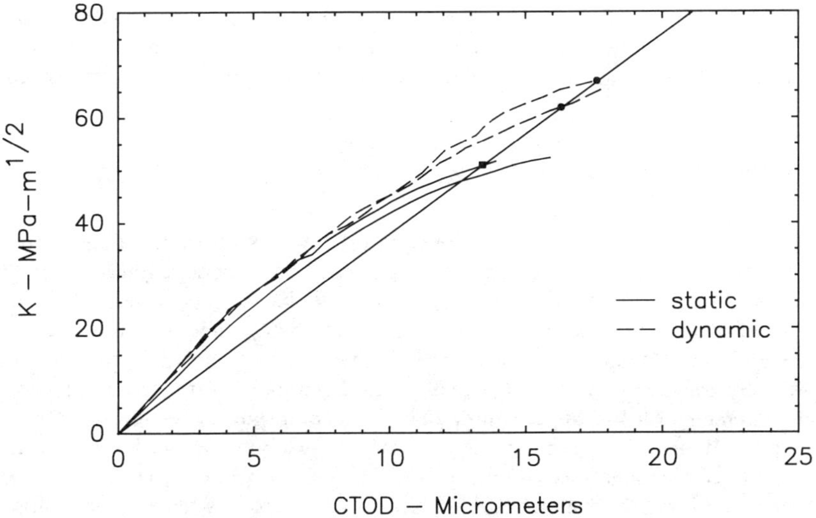

FIG. 10—*Static and dynamic* K *versus CTOD results for 4340 steel.*

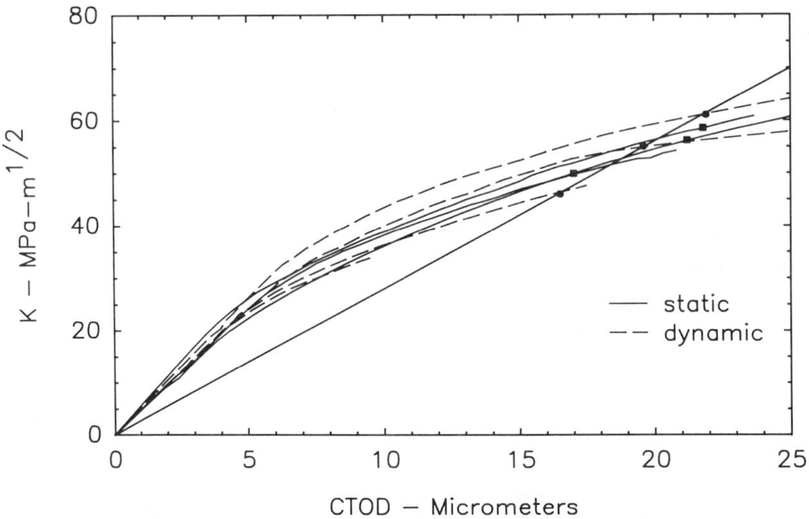

FIG. 11—*Static and dynamic* K *versus CTOD results for HY-100 steel.*

ing the straight line is evident. The shape of the *K*-CTOD curve is such that a small shift in K_q produces a large shift in the corresponding CTOD. Two of the dynamic CTOD plots are nearly identical, whereas the third is similar to the static plots. The HY-100 data of Fig. 11 show close agreement among the static tests, but larger scatter for the dynamic tests. Agreement between the two static results and the two dynamic results is much better for the tungsten data of Fig. 8 and the 4340 data of Fig. 10.

Intermediate rate tests were also run on these four materials [*1*] following the procedures in the appendix of ASTM Test E 399; the rates of those tests were approximately 10^3 MPa · $m^{1/2}$/s. A summary plot of the measured K_q for the four materials at the three rates is presented in Fig. 12. There is a lot of scatter at all rates, but the scatter in the dynamic results is not unreasonable. There is a trend for the SAE-01 steel to become more fracture-resistant; however, it is probably unwise to make a generalization about the other materials based on such limited data. But, it is safe to say that there is no dramatic change in fracture toughness with rate for any of the materials.

Discussion

As mentioned earlier, various assumptions are used in this method of determining the dynamic value of K_q. First, the stress intensity factor is assumed to increase linearly with time up to the onset of crack initiation. Second, the materials' elastic fracture response has to be time-independent, which ignores any viscoelastic effects. These hypotheses together allow the relationship between *K* and *t* to be quantified from the previously outlined companion specimen technique. Experimentally, the measured CTOD-time data show an initial linear increase, whereas separate strain measurements on the tup [*1*] show that the loading pulse rises linearly for a duration longer than the 10 or so microseconds required for a critical build-up of stress to occur. Furthermore, the finite-element analysis of Douglas [*17*] shows a nearly linear increase of load with time as does the analysis of Yokoyama and Kishida [*6*]. Therefore, the linear relationship between

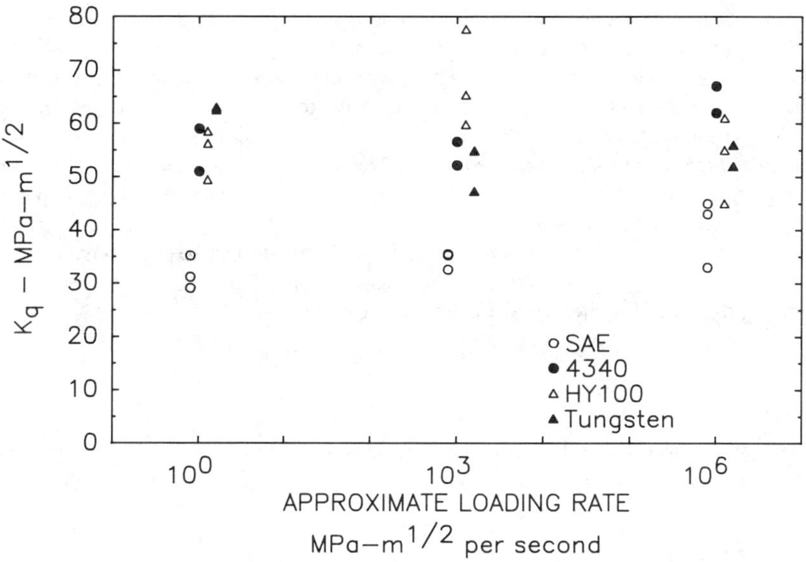

FIG. 12—*Static, intermediate-rate, and dynamic fracture toughness values for the four materials tested.*

time and the stress intensity factor seems valid at least up to crack initiation for the short duration fracture events measured here.

A more crucial assumption involves the validity of the reduced slope technique developed to predict dynamic crack initiation. This technique is modeled after the existing analytical method outlined in ASTM Test E 399. From this standard, if the quasi-static CMOD measurements of two geometrically similar specimens have the same elastic slope, a single reduced slope (or compliance) can be drawn that uniquely determines the onset of crack initiation for each specimen. Therefore, if geometrically similar quasi-static and dynamic specimens have the same elastic CTOD response, a single reduced slope should also determine the point of dynamic crack initiation. While this analogy is not perfect, the assumption seems like a logical first choice for empirically determining crack initiation using only the data present in this report.

An error analysis for the measurements and calculations of both the quasi-static and dynamic tests was completed in an earlier work [*1*]. These calculations showed that errors in the CTOD measurements were less than ±3% for both the static and dynamic tests. However, the relative uncertainty in determining the dynamic K_q is estimated to be ±20%. The main source of this large uncertainty is the determination of the elastic slopes of both the static and dynamic CTOD records; much of it comes from the variations among specimens. However, that is critical because the two elastic slopes are used to scale the dynamic load. Although the relative uncertainty is very large, the scatter in Fig. 12 is similar to that found by other investigators who have tested more than one specimen.

This measurement procedure could be considerably improved experimentally, and the data could be analyzed more rigorously. First, greater care in making the specimens as nearly identical as possible should reduce the scatter among "companion specimens." Second, generation of more data points in the elastic region to enable a better least-squares fit would yield more accurate elastic slopes; this can easily be done in the static tests and can be improved in the dynamic tests with a more powerful laser to produce a better signal-to-noise ratio. Also, the analysis could be conducted with the aid of a sophisticated finite element method (FEM) program to model the entire dynamic test. This would enable a hybrid experimental-numerical analysis that should produce more consistent results.

There are advantages to this procedure. The measurements are made at the crack tip, which is the region of interest and where wave effects do not have to be estimated. The specimen preparation and the experiments are fairly easy and inexpensive, which permits the testing of enough specimens to reduce the variability of the results. The measurements could be easily extended to lower or higher temperatures if needed. Also, the frequency response of the CTOD measurements is quite adequate.

Finally, although the dynamic K_q plotted in Fig. 12 are subject to the above concerns, the dynamic CTOD data are valid without any such assumptions. This information may be useful in understanding dynamic fracture and models that describe it.

Acknowledgments

This work was supported by Grant DAAG29-85-K-0218 from the Army Research Office.

References

[*1*] Sharpe, W. N., Jr., Douglas, A. S., and Shapiro, J. M., "Dynamic Fracture Toughness Evaluation by Measurement of CTOD," NTIS N88-22982/8, Army Research Office, Durham, NC, 1988.

[*2*] Shoemaker, A. K. and Seeley, R. S., "Summary Report of Round Robin Testing by the ASTM Group E24.01.06 on Rapid Loading Plane-Strain Fracture Toughness Testing," *Journal of Testing and Evaluation*, Vol. 11, 1983, pp. 261–272.

[*3*] Shoemaker, A. K. and Rolfe, S. T., "The Static and Dynamic Low-Temperature Crack-Toughness Performance of Seven Structural Steels," *Engineering Fracture Mechanics*, 1971, pp. 319–339.

[*4*] Rolfe, S. T. and Barsom, J. F., *Fatigue and Fracture Control in Structures: Applications of Fracture Mechanics*, Prentice-Hall, Englewood Cliffs, NJ, 1977.

[*5*] Giovanola, J. G., "Investigation and Application of the One-Point-Bend Impact Test," in *Fracture Mechanics: Seventeenth Volume, ASTM STP 905*, American Society for Testing and Materials, Philadelphia, 1986, pp. 307–328.

[*6*] Yokoyama, T. and Kishida, K., "A Novel Impact Three-point Bend Test Method for Determining Dynamic Fracture-initiation Toughness," *Experimental Mechanics*, Vol. 29, 1989, pp. 188–194.

[*7*] Costin, L. S., Duffy, J., and Freund, L. B., "Fracture Initiation in Metals Under Stress Wave Loading Conditions," in *Fast Fracture and Crack Arrest, ASTM STP 627*, American Society for Testing and Materials, Philadelphia, 1977, pp. 310–318.

[*8*] Chevallier, J. M., Ansart, J. P., and Dormeval, R., "Fracture Toughness of Some Metals Under High Loading Rate Conditions," *Institute of Physics Conference Series*, No. 70, 1984, pp. 229–236.

[*9*] Ravichandar, K. and Knauss, W. G., "Dynamic Crack-tip Stresses under Stress Wave Loading—a Comparison of Theory and Experiment," *International Journal of Fracture*, 1982, pp. 209–222.

[*10*] Rosakis, A. J., Duffy, J., and Freund, L. B., "The Determination of Dynamic Fracture Toughness of AISI 4340 Steel by the Shadow Spot Method," *Journal of Mechanics and Physics of Solids*, Vol. 32, No. 4, 1984, pp. 443–460.

[*11*] Kalthof, J. F., "Shadow Optical Method of Caustics," *Handbook on Experimental Mechanics*, Prentice Hall, Englewood Cliffs, NJ, 1987, Chap. 9.

[*12*] Epstein, J. S., Deason, V. A., and Reuter, W. G., "Dynamic Moire Interferometry Studies of Stress Wave and Crack-Tip Diffraction Events in 1018 Steel," in *Fracture Mechanics (19th Symposium), ASTM STP 969*, American Society for Testing and Materials, Philadelphia, 1988, pp. 482–503.

[*13*] Crostack, H. A. and Krüger, A., "Measuring Dynamic Stress Distributions by Laser Speckle Techniques," in *Impact Loading and Dynamic Behaviour of Materials Conference*, Bremen, 1987.

[*14*] Novinbakht, J., "Determination of Dynamic Stress Intensity Factor (K_I^{dyn}) Using Computer Vision," Ph.D. thesis, University of South Carolina, 1981.

[*15*] Demler, T. and Klenk, A., "Basic Research on the Near-Field of Dynamically Loaded Crack Tips," in *Impact Loading and Dynamic Behaviour of Materials Conference*, Bremen, 1987.

[*16*] Sharpe, W. N., Jr., "Applications of the Interferometric Strain/Displacement Gage," *Optical Engineering*, Vol. 21, 1982, pp. 483–488.

[17] Douglas, A. S. and Suh, M. S., "Impact Fracture of a Tough Ductile Steel," presented at the 21st Symposium on Fracture Mechanics, Annapolis, MD, 1988.

DISCUSSION

H. MacGillivray[1] (*written discussion*)—What is the maximum CTOD that the method could measure?

Tregoning et al. (*authors' closure*)—The limit was 25 μm.

Dr. Haressh Tippur[2] (*written discussion*)—How sensitive are your K^d measurements to the location of the indentations? What is the criterion behind the location of the indentation?

Tregoning et al. (*authors' closure*)—Yes, the K^d is very sensitive to the location of the indentations. We are locating the indentations 100 μm behind the crack tip. K^d measurement could be different if this is changed.

[1] Imperial College, Mechanical Engineering Department, Exhibition Road, London SW7 2A2, England.
[2] Postdoctoral research fellow, Graduate Aeronautical Laboratories, Caltech, Pasadena, CA.

Thomas Varga[1] *and Günther Schneeweiss*[1]

A New Method to Test Crack-Arrest Toughness by Using Three-Point Bend Specimens

REFERENCE: Varga, T. and Schneeweiss, G., **"A New Method to Test Crack-Arrest Toughness by Using Three-Point Bend Specimens,"** *Rapid Load Fracture Testing, ASTM STP 1130,* Ravinder Chona and William R. Corwin, Eds., American Society for Testing and Materials, Philadelphia, 1992, pp. 134–146.

ABSTRACT: This paper demonstrates the applicability of small and simple specimens to measure crack-arrest toughness, K_{Ia}. Specimens used up to now were large or of complicated geometry. Using small instrumented three-point bend specimens of simple geometry, possibilities and limitations were investigated. K_{Ia} values measured on compact crack arrest specimens and, respectively, by using three-point bend specimens have been compared. The results of such investigations on steel A 533 grade B class 1 are shown.

KEYWORDS: crack arrest, fracture toughness, three-point bend tests

In the following introduction, some typical cases are discussed, in which the assessment of safety against fracture is not possible without considering crack arrest.

1. The arrest of a running cleavage crack is of great importance in structures in which initiation may take place in local areas that have been embrittled during service.
2. Components subjected to temperature gradients may suffer crack initiation and subsequently propagation in the lower temperature areas. In the following, crack arrest may occur in an adjacent area of higher temperature, if adequate increase in toughness is present. To give an example: in the core belt region of a reactor pressure vessel (RPV), the combined effect of both phenomena, local embrittlement and temperature gradient, have to be considered.
3. Both crack initiation and crack-arrest behavior may also be observed at constant temperature. If two steels of different composition or microstructure, that is, of different toughness, are welded together, crack initiation may occur in the relatively brittle base material and arrest may follow in the tougher metal.
4. Crack initiation is also possible in a localized brittle zone (LBZ) of a weld heat-affected zone (HAZ) or in a weld metal itself. Only if arrest happens in an adjacent tougher microstructural area, is fitness for purpose confirmed. To give an example: the fracture surfaces of Charpy-V impact specimens and precracked Charpy-type specimens may exhibit multiple cleavage and arrest steps, especially in welds. Primary columnar grain and secondary fine grain microstructural areas were already compared in respect to their fracture tough-

[1] Both authors are with the Laboratory of Testing and Research in Material Technology (TVFA) and are professors at the Department of Materials Science and Testing, University of Technology Vienna, Karlsplatz 13, A-1040 Vienna, Austria.

ness. Plain carbon manganese steel weld metal was examined, while the notch position was parallel to the surface. A significant difference concerning the fracture toughness of the areas mentioned was found [*1*]. Crack arrest always was observed at the fine grain area, where higher toughness is present.

5. Finally, it has to be recorded that crack arrest is also achieved by a sufficient decrease of the stress intensity factor, K_I, along the course of a running crack.

Crack Arrest Testing Using CCA Specimens

The CCA specimen (crack-line-wedge-loaded compact-crack-arrest test specimen), as it is standardized in the ASTM Test for Determining the Plane-Strain Crack Arrest Fracture Toughness K_{Ia} of Ferritic Steels (E 1221), has been developed by the ASTM Crack Arrest Task Group (E 24.01.06). Within a round robin, three steels of different strength have been investigated. The K_{Ia} values of the reactor pressure vessel steel A 533 grade B class 1 serve as a reference in this paper.

In Ref *2*, the following properties are included: static yield strength at room temperature σ_{YS} = 480 N/mm^2; nil ductility transition temperature (*NDTT*) = −12°C; reference temperature *NDT* (ASME Boiler and Pressure Vessel Code, Section III, Division 1, NB 2331) RT_{NDT} = −2°C; and grain size according to ASTM No. 7-8. Additional data can be found in the ASTM Specification for Pressure Vessel Plates, Alloy Steel, Quenched and Tempered, Manganese-Molybdenum and Manganese-Molybdenum-Nickel (A 533/A533M-87): tensile strength R_m = 550 − 690 N/mm^2; percentage elongation after fracture $A(L_0$ = 50 mm) = 18% minimum; and Young's modulus E = 206 800 N/mm^2.

At the Laboratory for Testing and Research in Materials Technology (TVFA), Technical University Vienna, CCA specimens of the ASTM round robin with the designations KK-8, HH-6, E-7, Q-6, B-6, and WW-6 have been tested [*3*] according to ASTM Test E 1221. Dimensions were W = 160 mm, B = 50 mm, and B_N = 37.5 mm.

For hardfacing the welding rod, FOX DUR 500 was applied. Notch radii were 0.05 to 0.30 mm applied in steps of 0.05 or 0.1 mm. Testing temperature was 25°C for KK-8, HH-6, and E-7, and 10°C for Q-6, B-6, and WW-6.

Four specimens out of six showed crack arrest, whereas two broke into pieces. The results of the round robin, as published in Ref *2* are shown in Fig. 1. Results obtained at TVFA are also included [*4*].

Crack-Arrest Tests Using Three-Point Bend Specimens

The geometry of the specimen selected is shown in Fig. 2: W = width, B = specimen thickness, B_N = specimen thickness at crack plane. The span is designated as S. Some of the specimens were fatigue precracked with the initial notch radius being 0.05 mm and the depth of the precrack being larger than 5 mm.

Crack-mouth-opening displacement (CMOD) was measured according to the relative displacement of the knife edges near the front face of the specimen.

Because of the use of two types of specimens (with and without side grooves) and different displacement rates of the machine, 0.3 or 2 mm/min and 6 m/min, respectively, the following groups of tests were formed.

Series 1: Specimens without side grooves, quasi-static test.
Series 2: Specimens without side grooves, dynamic test.
Series 3: Specimens with side grooves, dynamic test.
Series 4: Specimens with side grooves, quasi-static test.

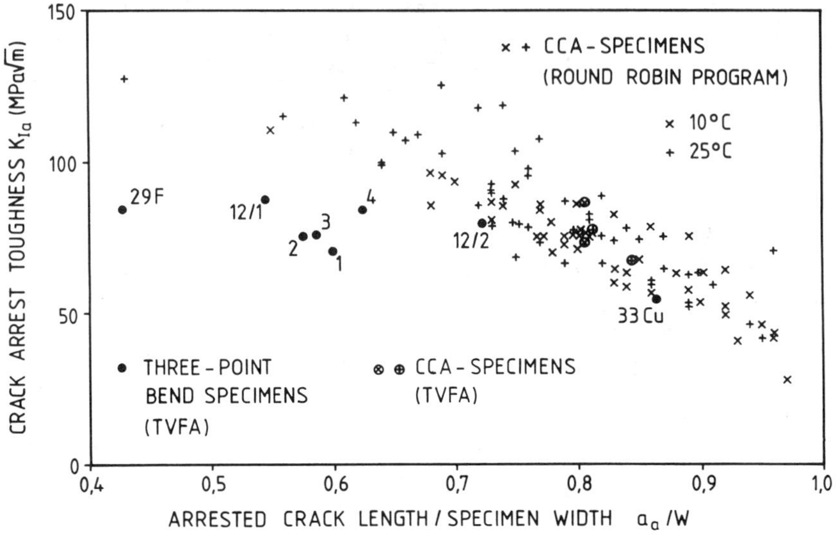

FIG. 1—*Crack-arrest toughness,* K_{Ia}, *as function of* a_a/W, *CCA specimens (round robin program and TVFA) and three-point bend specimens (TVFA), steel A 533 grade B class 1.*

In addition, the instrumentation and the conditions for crack initiation were varied to gain experience. The testing temperature was selected to achieve crack arrest with good probability for the steel and the specimen applied.

Test evaluation was performed by using the static arrest concept [5]. All those dynamic effects, which originate from the crack front and subsequently are reflected from the side wall to the tip of the running crack, are already extinct at the moment of crack arrest.

Both the force P and the CMOD measured almost immediately after crack arrest may be used to calculate the crack-arrest toughness, K_{Ia}.

The crack length, a_a, is defined according to the ASTM Test for J_{Ic}, a Measure of Fracture Toughness (E 813) as the mean value of the distance between the front surface of the specimen and the crack-arrest front on nine parallel lines.

$$a_a = \frac{1}{8}\left(\frac{a_0^* + a_8^*}{2} + \sum_{j=1}^{7} a_j^*\right) \tag{1}$$

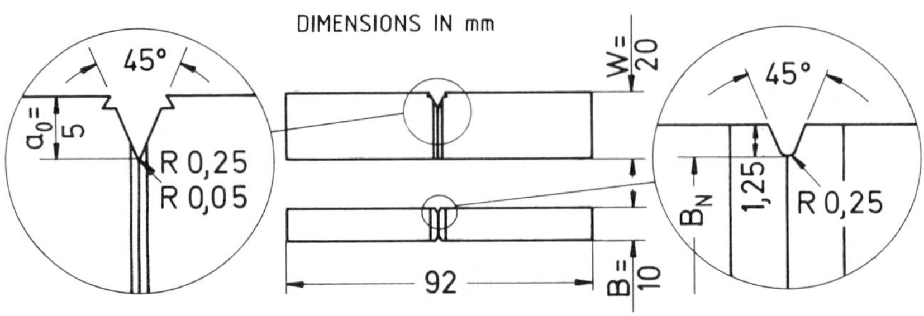

FIG. 2—*Dimensions of the three-point bend specimen. (If fatigue precracking was applied, the notch radius was 0.05 mm, and the depth of the precrack* a_0 *was larger than 5 mm.)*

The values K_{Ia} and CMOD may be calculated on the basis of the stress intensity factor and the maximum opening displacement observed on a crack of the length $2a$ as in a very large plate using the following formulas:

$$K_{Ia} = \sigma\sqrt{\pi a_a}\cdot f(x) = \frac{3P_aS}{2BW^2}\sqrt{\pi a_a}\cdot f(x) = K_{Ia}(P_a, a_a) \quad (2)$$

$$\text{CMOD}_R = \frac{4\sigma a_a}{E'}\cdot v(x) = \frac{6P_aSa_a}{BW^2E'}\cdot v(x) \quad (3)$$

where

$x = a_a/W$,

$E' = E/(1 - \nu^2)$ = Young's modulus in case of plain strain, and

ν = the ratio between transversal and longitudinal strain.

In case the ratio of S/W is about 4 (that is, the ratio between span and width of the specimen), the geometry factors $f(x)$ and $v(x)$ are defined according to Eqs 4 and 5; error limits for $f(x)$ are $\pm 0.5\%$ if $x \leqq 1.0$ [6] and for $v(x) \pm 1\%$ if $x \leqq 0.7$ [7].

$$f(x) = \frac{1.99 - x(1 - x)(2.15 - 3.93x + 2.7x^2)}{\sqrt{\pi}\,(1 + 2x)\,\sqrt{(1 - x)^3}} \quad (4)$$

$$v(x) = 1.45 - 2.18x + 13.71x^2 - 5.96x^3 - 36.9x^4 + 70.7x^5 \quad (5)$$

Series 1: Specimens Without Side Grooves, Quasi-static Tests

Eight tests were conducted, using the weld metal of a weld procedure test. The weld was prepared for a pipeline of 18-mm wall thickness and somewhat more than 1 m in diameter. The steel was St 480.7 TM according to DIN 17172.

The weld consumable was a high-strength welding rod, FOX CEL 90, with organic coating.

All of the eight three-point bend specimens were not side grooved. The initial fatigue precrack depth was between a_0 = 8.34 and 8.99 mm (a_0/W = 0.439 to 0.473). The following loading test was performed in a CMOD-testing device. The testing temperature was 0°C, achieved by using a Cryoson cooling chamber. The displacement rate was 0.3 mm/min.

The force-opening displacement diagrams were directly registered by using an *xy* recorder; Fig. 3 gives the diagram of the Specimen 291 as an example [4].

In addition, force and opening displacement were recorded in function of time by a Nicolet Explorer III (2090-3B) digital storage oscilloscope, using a measuring interval of 0.5 s. At the crack jump, only the two values indicated by small circles in Fig. 3 have been stored. Between these two points, the dotted line is due to the characteristics of the *xy* recorder.

The following observations could be made:

1. Before and after crack jump, the chosen displacement rate of the loading point of 0.3 mm/min corresponds to a crack-mouth opening displacement velocity of less than 0.24 mm/min = 0.004 mm/s; the applied recording interval of 0.5 s results in an opening of less than 0.002 mm. This is, therefore, the upper limit of possible error in recording CMOD as a result of the measuring interval. If one does not accept this error value, then the measuring interval is to be reduced. Unfortunately, a similar estimation of the error is not possible for P_a.

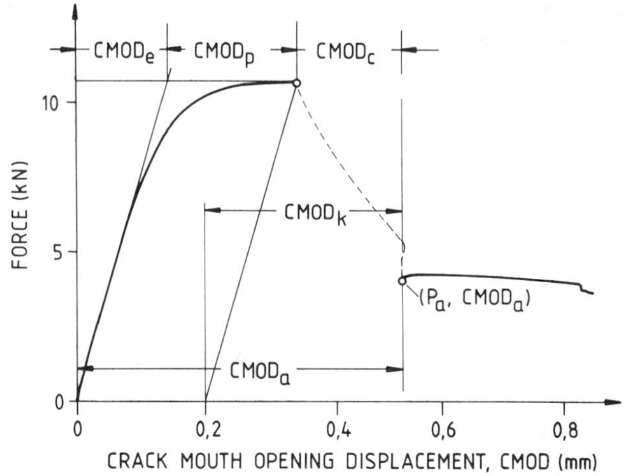

FIG. 3—*Force in function of CMOD, Specimen 291 (Series 1).*

2. Specimens without side grooves as used in Series 1 may exhibit an unacceptably tunneled crack front, as can be seen in Fig. 4, Specimen 155 [*4*].
3. Some specimens exhibit too large a plastic deformation before the actual crack starts. It is assumed that the stress distribution in the ligament at crack arrest will not be altered essentially by the nonlinear part of CMOD (= nonproportional CMOD = $CMOD_p$) at the start of the crack jump.

Hence, a correction is necessary:

$$CMOD_k = CMOD_a - CMOD_p = CMOD_e + CMOD_c \tag{6}$$

That is, the overall opening $CMOD_a$ is reduced by $CMOD_p$ (see Fig. 3). The value $CMOD_e$ is the linear part of CMOD at the crack jump start, and $CMOD_c$ is caused by the crack jump.

FIG. 4—*Fracture surfaces, specimens 155 (Series 1), 2 (Series 2), 12 (Series 3), and 29F (Series 4). The suffix (S) indicates side grooving.*

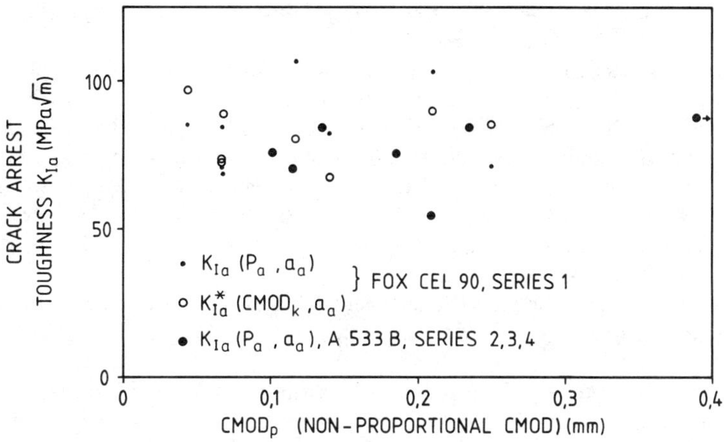

FIG. 5—K_{Ia} *derived by using* P_a *and* a_a *and* K_{Ia} *derived by using* $CMOD_k$ *and* a_a *over the nonproportional CMOD. There is no relation of any kind to the nonproportional CMOD. Steels investigated are: weld metal FOX CEL 90 and RPV-steel A 533 grade B class 1.*

Based on the measured values of P_a, a_a, and CMOD$_k$ of Series 1, K_{Ia} in function of P_a and a_a according to Eq 2 and K^*_{Ia} in function of CMOD$_k$ and a_a (as given in Eq 7) may be calculated.

$$K^*_{Ia}(\text{CMOD}_k, a_a) = \frac{\text{CMOD}_k}{\sqrt{a_a}} \cdot \frac{E'f(x)\sqrt{\pi}}{4v(x)} \tag{7}$$

Equation 7 results by eliminating P_a from Eqs 2 and 3.

In Fig. 5, K_{Ia} and K^*_{Ia} are given in dependence of CMOD$_p$ (as well as K_{Ia} for Series 2, 3, and 4, according to Table 1). As it is demonstrated, the arrest fracture toughness is practically not influenced by the nonproportional part of CMOD, that is, CMOD$_p$. K_{Ia} values calculated by using CMOD$_k$ are therefore equivalent to those measured on specimens behaving in a linear manner before the cleavage crack initiation. Linear behavior is generally regarded as appropriate.

The mean value of CMOD$_k$, 0.277 mm, based on the measured CMOD$_a$ and CMOD$_p$ values, is in good agreement with the mean calculated value of CMOD$_R$ for elastic behavior according to Eq 3, 0.283 mm, derived from the measured values P_a and a_a.

Both results confirm that deduction of CMOD$_p$ from CMOD$_a$ is justified. This deduction is also applied in Series 2, 3, and 4. The stepwise straining of the specimen to reduce the nonlinear part of CMOD as mentioned in ASTM Test E 1221 for the CCA specimen was not applied.

Series 2: Specimens Without Side Grooves, Dynamic Tests

The specimens of Series 2, 3, and 4 were cut from the broken halves of the CCA specimens. Therefore, the K_{Ia} values measured can be compared to those obtained by using the CCA specimens.

The testing machine used for Series 2, 3, and 4 was a pendulum by Schnadt, of the Type SPH-III (produced by Mohr & Federhaff AG), equipped with a variable loading rate device and a working capacity equal to 100 J. The span always was $S = 83$ mm (the free distance between the anvils being 80 mm, and the radius 1.5 mm).

The displacement rate in testing Series 2 was 6 m/min. Testing temperature of the four specimens, designated 1 through 4, was 0°C.

The force was measured by strain gages attached to the loading chisel. The displacement of the chisel, that is, the loading point displacement, was measured by using a linear variable displacement transformer (LVDT); the crack opening was obtained by using a displacement measuring device equipped with strain gages. The electric potential was measured by appropriate means without using an active current [*8*]. After amplifying the signals, they were fed to a four-channel multiplexer and subsequently linked to a channel of the digital storage oscilloscope. The measuring sequence was 0.02 ms (0.01 ms for Specimen 2). The curves were drawn by connecting the measured points using straight lines.

The electric potential, the opening displacement, CMOD, and the displacement of the loading point in function of time are shown in Fig. 6 for Specimen 1. (The position of the point of origin is of no significance.)

The measuring equipment is discussed in the description of Series 4. Contrary to Series 4 for the measurement of the loading point displacement, an amplifier was used whose upper cutoff frequency limit was 0.5 kHz (−1 dB), the phase delay was 570 μs, and the impulse rise time was 600 μs. Therefore, higher frequency vibrations were not registered; furthermore, the signals were delayed.

The minimum registered value of P at the end of the crack jump is designated as P_{cL} (see Fig. 6). For calculating K_{Ia} according to Eq 2, the deterministic value of P_a has to be selected along the curve.

Here, not P_{cL}, but the mean value over the period of 2 ms after the crack jump was defined as P_a, the crack-arrest force. The period began with P_{cL}, and 25 measured points including P_{cL} were used for calculating the mean value (except for Specimen 2 with 50 measured points). In general, there was not much systematic variation of the force during this time period. The calculated K_{Ia} values are included in Table 1 and in Fig. 1.

The crack-mouth-opening displacement immediately increased after the crack jump. To obtain $CMOD_c$ caused by the crack jump, the following procedure was applied: CMOD in function of time was divided into two parts: one before and the other after the crack jump. For each of the parts, the mean lines were drawn. The difference of the opening displacement at the time of

TABLE 1—*Crack arrest tests using three-point bend specimens, steel ASTM A 533 grade B class 1.*

	Series (Side Grooves)							
	2 (Without)				3 (With)			4 (With)
	Specimen							
Characteristics	1	2	3	4	12/1	12/2	33 Cu	29 F
a_0/W	0.460	0.442	0.438	0.437	0.421	. . .	0.253	0.200
$CMOD_p$ (mm)	0.116	0.186	0.102	0.135	0.707	. . .	0.210	0.236
$CMOD_k$ (mm)	0.262	0.231	0.267	0.244	0.214	. . .	0.622	0.255
a_a (mm)	12.25	11.56	11.72	12.49	10.90	14.64	17.26	8.54
a_a/W	0.613	0.583	0.591	0.625	0.545	0.732	0.863	0.427
P_{cH} (N)	9917	10 437	10 615	10 484	10 681	7335	14 589	17 680
P_{cL} (N)	4911	7 173	5 935	6 824	6 364	2495	516	9 380
P_a (N)	6037	7 156	7 036	6 907	7 265	2917	731	10 110
K_{Ia}, MPa $m^{1/2}$	70.17	75.23	75.93	84.24	87.78	79.22	54.74	84.34
Temperature, °C	±0	±0	±0	±0	+28	+28	±0	+20
Displacement rate	6 m/min				6 m/min			2 mm/min

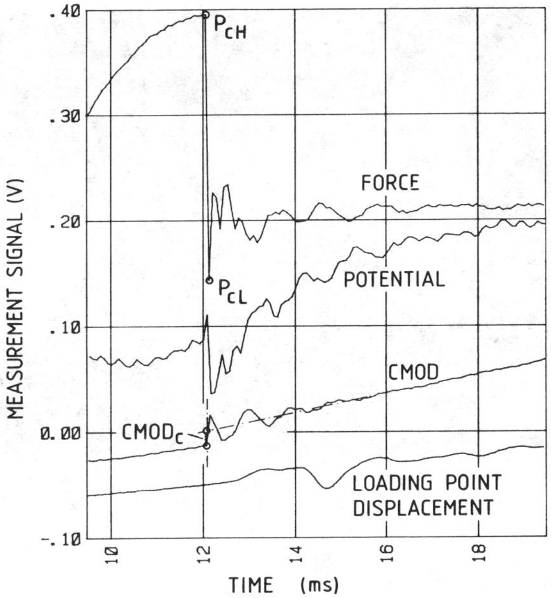

FIG. 6—*Force, electric potential, crack-mouth opening displacement, and loading point displacement in function of time, Specimen 1, Series 2.*

crack jump was taken as $CMOD_c$ (see Fig. 6). Therefore, the corrected crack-mouth-opening displacement was $CMOD_k = CMOD_e + CMOD_c$.

Corresponding to Series 1, a strong curvature of the crack front was observed after arrest.

Because of the increased bending velocity, an additional influence has to be regarded: under dynamic loading, the specimen will be broken, in spite of possible multiple arrest steps along the crack path, under a single stroke.

Therefore, there will be a less pronounced crack-arrest trace on the crack surface, as Specimen 12 in Fig. 4 shows. Similar to Series 1, there was no or very little crack propagation on the side of the specimen at the first jump. It is most probable that the first crack jump separates only the center part of the specimen below the notch by cleavage, whereas the volume near the sides of the specimen is not yet separated. Another difficulty in measuring the crack jump length is that the borderlines are sometimes difficult to observe, at least at usual magnification. However, crack arrest was usually well detectable in the scanning electron microscope (Fig. 7).

The determination of P_a and, above all, of $CMOD_c$ is less certain as in a quasi-static test. Regarding the test temperature of 0°C and the displacement rate of 6 m/min, the measured K_{Ia} values are lower than the corresponding CCA values and lie in the expected range (as far as can be assessed).

Series 3: Specimens with Side Grooves, Dynamic Loading

The specimens tested, marked as 12 and 33Cu, were both of the dimension $W = 20.00$ mm, $B = 10.00$ mm, and $B_N = 7.50$ mm.

Specimen 12 was precracked to a depth of $a_o = 8.42$ mm. The notch of specimen 33Cu was machined according to Fig. 2 (depth 5 mm, radius 0.25 mm), but with an additional notch, the width being 0.4 mm, the radius 0.2 mm, and the largest depth to the surface 5.25 mm. In this additional notch, a copper wire of 0.2-mm diameter was laid and melted by an electron beam. Testing was conducted similar to Series 2; testing temperature of Specimen 33Cu was 0°C; that of Specimen 12 was +28°C.

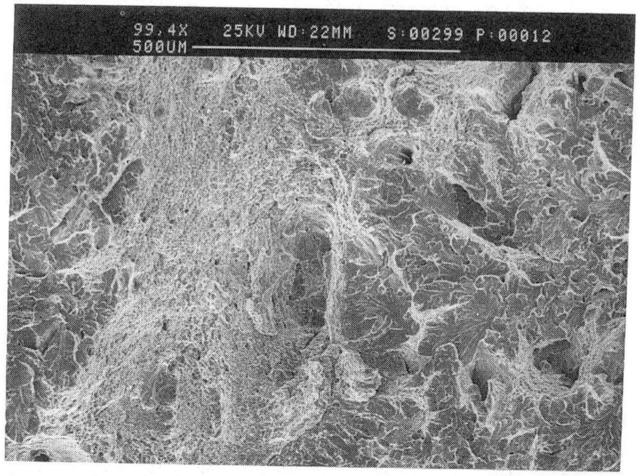

FIG. 7—*Fracture regions along the crack path in Specimen 12, Series 3: cleavage fracture (first jump)-dimple fracture (arrest)-cleavage fracture. Crack propagation was from the right to the left.*

To give another example, the force, the displacement of the loading point, and the electric potential, all in function of time, are shown in Fig. 8 as measured on Specimen 12. Four crack jumps were observed, which were indicated also in the potential curve. The first two crack jumps were evaluated (see 12/1 and 12/2 in Table 1). The fracture surface of this specimen is depicted in Fig. 4.

Specimen 33Cu showed a very low force value after the arrest of a big crack jump. Strong

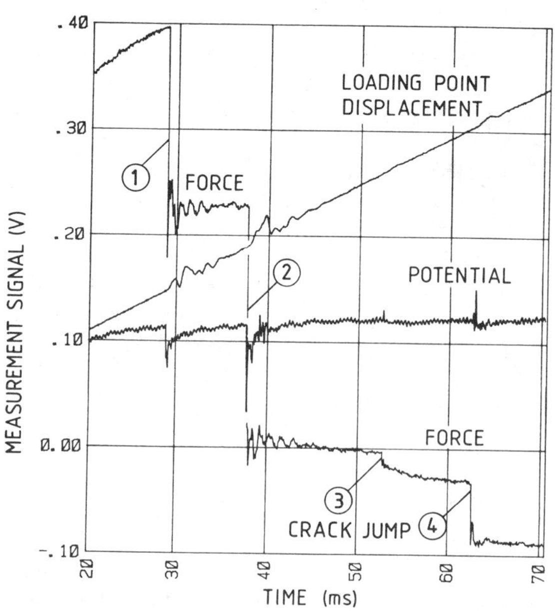

FIG. 8—*Force, loading point displacement, and electric potential in function of time, Specimen 12, Series 3.*

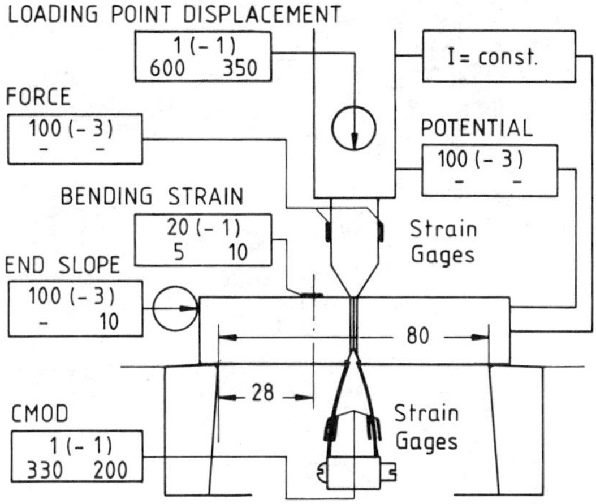

FIG. 9—*Measuring channels as applied in Series 4; indicated are the upper cutoff frequency in kilohertz, amplitude drop in decibels given in parentheses, phase delay in microseconds, and impulse rise time in microseconds.*

vibrations were observed. The crack-arrest toughness K_{Ia} was only 54.7 MPa $\cdot$ m$^{1/2}$ and is included in Fig. 1.

Some conclusions for this group of tests: unlike Series 2, side-grooved specimens were examined in Series 3 and one of the specimens, 33Cu, had a copper embrittled notch. Because of the relatively elevated constraint near the side surfaces of the grooved specimens, the crack front deviates far less from a straight line than those of Series 1 and 2, which were not side-grooved. Therefore, multiple crack jumps are easy to evaluate. The copper wire, melted into the notch, was of no advantage in this case.

Series 4: Specimens with Side Grooves, Quasi-static Test, High Recording Velocity

The measuring device for dynamic testing in Series 3 had the disadvantage of low storage capacity, because of the oscilloscope used. Only 4000 points could be stored in one channel.

Because of the use of a 4-channel multiplexer, each measuring channel had only 1000 points at disposal. The smaller the time per point is, the shorter the time for recording becomes. This was tolerable for dynamic testing, but for quasi-static testing a higher storage capacity was judged to be necessary.

In Series 4, an Ampex PR 2230 14-channel magnetic tape recorder was used with a cutoff frequency of 80 kHz (Wideband Group I according to IRIG-Doc. 106) at the highest recording velocity of 120 ips. If playback velocity is lower (the smallest one being 15/8 ips), the signals at crack jump may be analyzed. A Gould ES 1000 high-speed recorder was applied for graphic display.

The measuring equipment is depicted in Fig. 9. The connection of the magnetic tape recorder to the device used for graphic display is not shown. Every amplifier used in the measuring channel (as shown in Fig. 9) is labeled by the upper cutoff frequency in kilohertz (1, 20, or 100), according to a drop in amplitude, given in decibels (−1 or −3), phase delay in microseconds, and the impulse rise time in microseconds. The latter are taken from the manufacturers' manuals.

In addition to the measured characteristics in Series 1, 2, and 3 (force, loading point displacement, and opening displacement in function of time), a linear variable differential transformer (LVDT) was attached to the end surface of the specimen. This LVDT indicated the angle of the deflection line, and the deflection itself could be estimated.

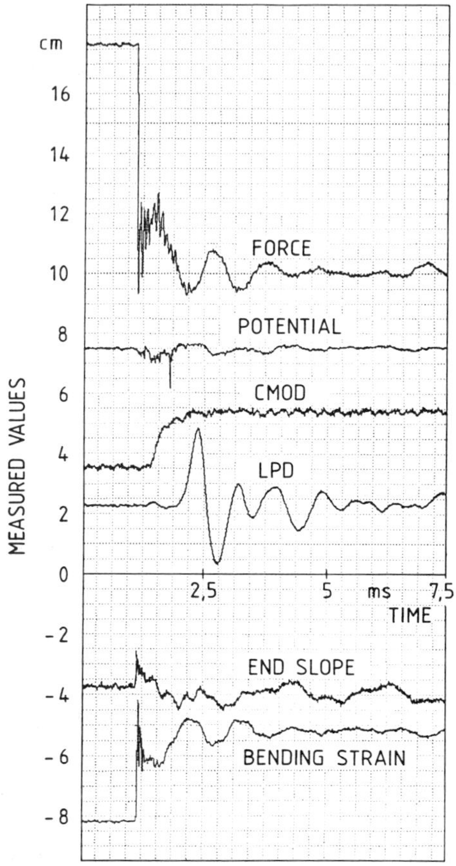

FIG. 10—*Measured values in function of time, Specimen 29F, Series 4. Scales are: 1 cm on the vertical axis corresponds to the force of 1 kN; to 0.2 V of electric potential (between chisel and specimen); to 0.1 mm of CMOD; to 0.5 mm of load point displacement (LPD); and to 200 μm/m bending strain (end slope noncalibrated).*

A strain gage was attached to the compressive side of the specimens 29.5 mm from one anvil. This was used to check whether the bending stress in the specimen was increasing at the same rate as the force measured on the chisel.

In a Charpy impact test, immediately after the cleavage crack jump, Beisken [*9*] observed a short-term loss of contact between the specimen and the tup of the pendulum. To check whether this effect is also present in static tests, a constant current source was positioned between the chisel and the specimen. A drop of potential indicated the short decrease of the contact force. Hence, the potential measurement is different in this case from that in Series 2 and 3.

To minimize the influence of contact resistances, both the chisel and the specimen had a separate feeding in of the current and separate contacts for potential measurements.

Four specimens were tested at ambient: Specimen 6 was fatigue precracked; three specimens, 8F, 29F, and 30F, were hard faced by using the welding rod FOX DUR 500 and notched (with a_0 = 4 mm, a_0/W = 0.200, notch radius 0.25 mm) but not fatigue precracked.

Specimens 29F and 30F were side grooved. In Specimens 6 and 8F, no crack jump occurred. Specimen 30F broke through with no crack arrest. Specimen 29F showed two cleavage crack jumps; the first of these was evaluated. The measured characteristics are depicted in Fig. 10 in function of time.

As expected from the characteristics of the measuring channels, the curves showing the force, the bending angle (end slope), the potential change, and the bending strain do react immediately to the crack jump (see Fig. 10). On the contrary, the displacement of the loading point and the opening displacement indicate only lower frequency vibrations that could be registered. Additionally, the signals are displayed with delay. For conducting a more precise analysis, the inertia momentum of the chisel on the one hand, and the self frequencies of the measuring device and of the measuring channels on the other hand, have to be taken into account. Moreover, Fig. 10 demonstrates that after 2 ms the opening displacement has stabilized.

A sudden drop of the force measured on the chisel appeared at the moment of the crack jump: it decreased from P_{cH} = 17.68 kN to P_{cL} = 9.33 kN. Subsequently, the force was quickly recovering and it stabilized after a strongly damped oscillation of about 4 ms at 10.11 kN. The latter was taken as the arrest force P_a. The corresponding value of K_{Ia} is given in Table 1 and shown in Fig. 1. The fracture surface is depicted in Fig. 4.

Above all, Series 4 was needed for testing an improved instrumentation. This instrumentation turned out to be satisfactory for the measurements envisaged. For future tests, it is intended to apply measuring channels that all use the same amplifiers or at least amplifiers with similar characteristics and especially with high measuring frequencies.

The application of a brittle crack starter weld seems to be of great advantage. The best way of producing such a weld is currently under investigation.

Conclusions

According to the investigations performed, the three-point bend specimen as selected with side grooves is applicable for crack-arrest toughness, K_{Ia}, measurements. Best experience was gained by using quasi-static loading. Size requirements have to be clarified by further testing.

The amount of plastic deformation before cleavage crack initiation did not cause any significant changes concerning the crack-arrest toughness value. By omitting the nonlinear part, $CMOD_p$, of CMOD determined in the force-CMOD diagrams at crack jump start, the other part can be used to evaluate results of precracked Charpy-type or even of Charpy-V specimens to obtain crack-arrest toughness, K_{Ia}. More work is necessary to investigate how the K_{Ia} values, which are derived from small specimens, may yield representative data for large sections.

References

[1] Konkoly, T., Straube, H., and Varga, T., "Investigations on MAG Weld Metal for Critical Valuation of Fracture Mechanics Properties," in *Advances in Fracture Research, Proceedings of the 6th International Conference on Fracture (ICF 6)*, Vol. 2, S. R. Valluri, D. M. R. Taplin, P. Rama Rao, J. F. Knott, and R. Dubey, Eds., Pergamon Press, Oxford, 1984, pp. 1137–1143.

[2] Barker, D. B., Chona, R., Fourney, W. L., and Irwin, G. R., "A Report on the Round Robin Program Conducted to Evaluate the Proposed ASTM Standard Test Method for Determining the Plane Strain Crack Arrest Fracture Toughness, K_{Ia}, of Ferritic Materials," U. S. Nuclear Regulatory Commission NUREG/CR-4996 (Oak Ridge National Laboratory ORNL/Sub/79-7778/4), University of Maryland, College Park, MD, 1988.

[3] Kanada, A. H., "Crack Arrest Investigations on Weldable Steels (Rißauffanguntersuchungen an Schweißbaren Stählen)," thesis, University of Technology Vienna, (TU Wien), 1987.

[4] Schneeweiss, G. and Varga, T., "Experience Gained by Using Three-Point Bend Test Specimens to Investigate Crack Arrest Toughness K_{Ia} of a Low-Alloy Steel (Erfahrungen mit Dreipunkt Biegeproben zur Ermittlung der Rißauffangzähigkeit K_{Ia} an Niedriglegiertem Stahl)," *Swiss Materials*, Vol. 1, No. 6, Dec. 1989, pp. 29–37.

[5] Ripling, E. J., Crosley, P. B., and Wiersma, S. J., "A Review of Static Crack Arrest Concepts," *Engineering Fracture Mechanics*, Vol. 23, No. 1, 1986, pp. 21–33.

[6] Srawley, J. E., "Wide Range Stress Intensity Factor Expressions for ASTM E 399 Standard Fracture Toughness Specimens," *International Journal of Fracture Mechanics*, Vol. 12, No. 6, 1976, pp. 475–476.

[7] Nisitani, H., Mori, K., and Noguchi, H., "An Analysis of Single-Edge-Cracked Specimen under Three- or Four-Point Bending by Body Force Doublet Method," *Transactions of the Japanese Society of Mechanical Engineers*, Vol. 52, No. 474, 1986, pp. 539–543.
[8] Loibnegger, F., Salzmann, F., and Varga, T., "Detection of Crack Initiation in Fatigue Precracked Charpy-type Specimens," *Nuclear Engineering and Design*, Vol. 96, 1986, pp. 139–148.
[9] Beisken, H., "Fracture Mechanics Investigations Using Impact Bend Specimens (Bruchmechanische Untersuchung mittels Kerbschlagproben)," Thesis 5114, Swiss Federal Institute of Technology (ETH), Zurich, 1973.

DISCUSSION

Randy K. Nanstad[1] *(written discussion)*—You indicated that you performed impact tests at 5 m/s as well as at 0.1 m/s and slower. Have you performed similar analysis for those faster tests, and if so, how did you resolve the load-time traces given your stated rise time of 200 μs, which I believe is too long for tests at 5 m/s?

Thomas Varga (author's closure)—To achieve satisfactory load-time traces for impact tests at 5 m/s, the rise time of 200 μs has to be reduced substantially, which means that the upper cutoff frequency of all components of the measuring channel has to be increased.

[1] Oak Ridge National Laboratory, Oak Ridge, Tennessee.

John H. Underwood,[1] I. A. Burch,[2] and J. C. Ritter[2]

Crack-Arrest and Static Fracture Toughness Tests of a Ship Plate Steel

REFERENCE: Underwood, J. H., Burch, I. A., and Ritter, J. C., **"Crack-Arrest and Static Fracture Toughness Tests of a Ship Plate Steel,"** *Rapid Load Fracture Testing, ASTM STP 1130,* Ravinder Chona and William R. Corwin, Eds., American Society for Testing and Materials, Philadelphia, 1992, pp. 147–160.

ABSTRACT: The recently standardized ASTM Test for Determining the Plane-Strain Crack Arrest Fracture Toughness K_{Ia} of Ferritic Steels (E 1221) was modified for use with BIS 690 steel, an Australian low-alloy, 700-MPa strength, quenched and tempered steel used for ship plate. Specimens of 50-mm thickness in longitudinal and transverse orientations were tested at −60°C with various depths of side groove. Specimen configurations somewhat outside the range of ASTM Test E 1221 were found to be useful for this steel and were suggested as modifications to the standard method. A different solution and expression for stress intensity factor were used to evaluate the tests and to develop a method for predicting the crack length at which a running crack will arrest. Good agreement was obtained between the observed and predicted crack lengths at arrest.

The wedge-loading arrangement of the crack-arrest test procedure was used for static fracture toughness tests. Procedures for static wedge-loading fracture toughness were proposed, particularly for comparisons with crack-arrest toughness test results obtained under similar test conditions. For the tests here, the static fracture toughness was more than twice the value of crack-arrest toughness.

KEYWORDS: steel, crack arrest, fracture toughness, ship plate, *K* expression

Objectives

Fracture toughness for rapid load conditions is a common concern with military structures. A long-standing procedure for addressing this concern in ship structures is the explosion bulge test developed by the U.S. Navy [*1*]. This procedure gives an effective simulation of explosive loading conditions of ship plate and has very successfully predicted service behavior. For armament components such as cannons, full-scale fatigue tests using rapid load firing tests have been shown by the U.S. Army to give excellent predictions of service behavior [*2*]. Regardless of how well such full-scale tests can predict service behavior for these components, there are significant drawbacks. Full-scale tests are always costly in time and money, and the test results often cannot be related directly to accepted fracture mechanics properties and analysis, thus requiring additional tests when conditions change.

The ASTM Test for Determining the Plane-Strain Crack Arrest Fracture Toughness K_{Ia} of Ferritic Steels (E 1221) may provide a means to relate directly and quantitatively the rapid load fracture behavior of components such as ship plate and cannon to a material fracture property. The K_{Ia} method has the advantages of being a reasonably small laboratory test and one whose

[1] Research engineer, Benet Laboratories, Army Armament Research, Development and Engineering Center, Watervliet, NY 12189-4050.

[2] Experimental officer and senior principal research scientist, Materials Research Laboratory, Defence Science and Technology Organisation, Melbourne, 3032, Australia.

TABLE 1—*Material composition.*

Composition, Weight Percent									
C	Mn	Si	Ni	Cr	Mo	B	Ti	Nb	V
0.16	1.5	0.40	0.25	0.35	0.40	0.005	0.05	0.05	0.09

results can be related directly to the level of applied stress intensity factor of a loaded component. Therefore, the K_{Ia} test can be used in the same general way that plane-strain fracture toughness, K_{Ic}, is used, that is, as a critical material property for use with fracture mechanics analysis to predict the load and geometry conditions at which fracture will occur.

The overall objective of the work here was to demonstrate that K_{Ia} gives a consistent laboratory measure of the crack-arrest fracture toughness property of a ship plate steel for various test configurations. Starting notch length and depth of side grooves were chosen as configurational variables in tests with specimens half the depth, W, of that recommended in ASTM Test E 1221. If consistent results could be obtained with smaller specimens, the test would become a more practical laboratory procedure. The smaller specimen would allow more choice of test location, such as around welds in ship plate or at different locations and orientations in cannon components. Rosenfield et al. [*3*] showed that miniature specimens for reactor surveillance testing gave slightly lower K_{Ia} values than larger specimens from an ASTM A508 steel.

As the investigation proceeded, some progress was noted in the use of test configurations outside the recommended range and the use of analysis to predict conditions for a successful arrest of a running crack. Therefore, a second objective of the investigation became the development of modified K_{Ia} test and analysis procedures for this steel and their proposed use for K_{Ia} testing in general.

Material and Test Procedures

Material

The steel used for the tests was Australian BIS 690, a 50-mm-thick, low-alloy ship plate steel, quenched and tempered to a nominal 700-MPa strength. The chemical composition and mechanical properties of the plate from which all specimens were taken are given in Tables 1 and 2, respectively. The initial plan was to perform the K_{Ia} tests at −40°C, a typical extreme low service temperature for military hardware. Note that the Charpy energy in Table 2 at −40°C is considerably above the general recommendation of 41 J for a successful K_{Ia} test [*3*]. This recommendation proved to be good advice; all tests but one were performed at a lower temperature, −60°C, to obtain successful crack arrests.

The room temperature fracture toughness of the material was characterized by J_{Ic} tests of

TABLE 2—*Mechanical Properties.*

Yield Strength, L, +20°C, MPa	Tensile Strength, L ,+20°C, MPa	Fracture Toughness, from J_{Ic} Tests, T-L, +20°C, MPa · m$^{1/2}$	Charpy Energy, Joules, T-L, Temperature, °C			
			0	−20	−40	−50
701	762	239	91	87	78	60

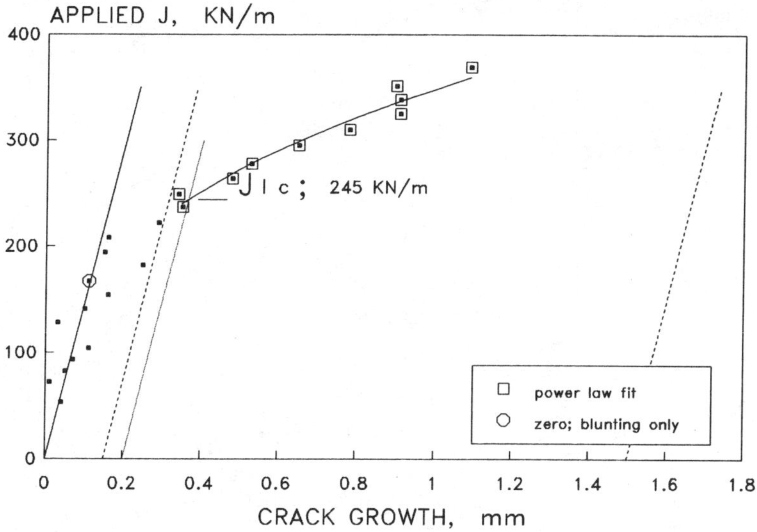

FIG. 1—*Applied* J *versus crack growth for BIS 690 at +20°C.*

25-mm-thick compact specimens, which yielded a mean J_{Ic} value of 251 KN/m and corresponding K value of 239 MPa · $m^{1/2}$. One of the two J versus Δa plots is shown as Fig. 1. Unloading compliance was used to determine Δa, following the ASTM Test for J_{Ic}, a Measure of Fracture Toughness (E 813) with one addition. A single point at a J value of about two thirds of the expected J_{Ic} was used as a reference point. Shifting this point (and all data) to fall exactly on the blunting line involved the use of an effective elastic modulus, E, of 209.7 GPa, compared with the initial nominal value of 210.0 GPa. For the tests here, this reference point technique was effective in addressing the common problem in J_{Ic} tests of Δa shifts of the data.

K_{Ia} *Tests*

The crack-arrest test specimens (Fig. 2) followed the recommendations of ASTM Test E 1221 except for two modifications. First, the side groove depth was varied from $B_N/B = 0.75$ (recommended) to $B_N/B = 1$. Second, the initial notch was varied from $a_0/W = 0.30$ (the minimum recommended) to $a_0/W = 0.16$. The width-to-thickness ratio, W/B, was 2.0, which, although not outside the recommended range, $2.0 \leq W/B \leq 8.0$, was half the value commonly used. For a given plate thickness, a specimen with $W/B = 2$ is less likely to result in arrest than a specimen with larger face dimension. However, the $W/B = 2$ specimen would be more useful, as discussed earlier, so it is worth pursuing.

A brittle weld was added at the notch tip (using Hardex N electrodes), and the wedge-load-type tests of T-L and L-T orientations were performed generally at −60°C. The low temperature was attained by pumping a liquid and gas mixture of nitrogen into a foam-plastic enclosure around the specimen. The side groove and initial notch conditions and the initial displacement, δ_0, and resulting initial applied stress intensity, K_0, for the tests are listed in Table 3. The expression used for calculating the ratio K/δ as a function of a/W is different from that in ASTM Test E 1221, and is the following [*4*]:

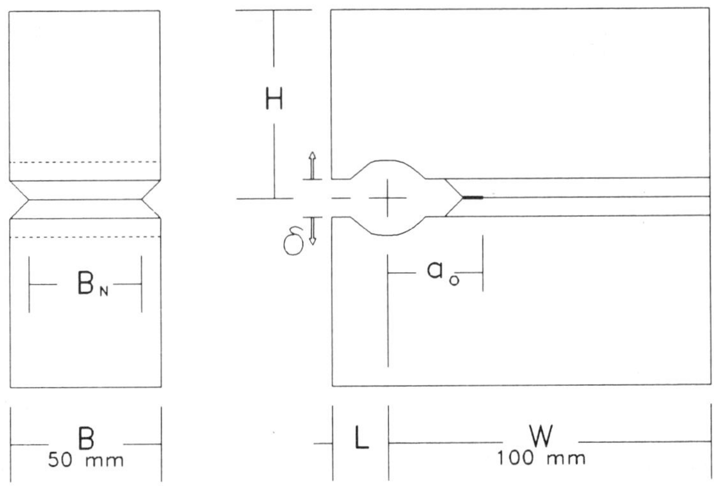

FIG. 2—*Specimen configuration for* K_{Ia} *tests.*

$$KW^{1/2}/\delta E(1 - a/W)^{1/2} = 0.748 - 2.176(a/W) + 3.56(a/W)^2 - 2.55(a/W)^3 + 0.62(a/W)^4$$

$$\text{for } 0.2 \leq a/W \leq 1.0 \tag{1}$$

This expression gives similar results to that in ASTM Test E 1221 for $0.4 \leq a/W \leq 0.6$, but differs elsewhere. For $a/W = 0.8$, a crack length often used in K_{Ia} tests, Eq 1 gives a value 7% higher than the ASTM Test E 1221 expression. The ASTM Test E 1221 expression is repeated here for reference, as follows:

$$KW^{1/2}/\delta E = \frac{2.24[1.72 - 0.9(a/W) + (a/W)^2](1 - a/W)^{1/2}}{[9.85 - 0.17(a/W) + 11(a/W)^2]}$$

$$\text{for } 0.30 \leq a/W \leq 0.85 \tag{2}$$

A comparison of the K/δ expression from collocation analysis [*4*], Eq 1, with that from ASTM Test E 1221 based on experimental compliance tests, Eq 2, is shown in Fig. 3. Pajot's recent finite element results for the same wedge-load compact configuration [*5*] are also shown. The two independent sets of analytical results agree within 2% for a/W up to 0.5 and within 1% for $0.5 \leq a/W \leq 0.95$. Both sets of analytical results agree well, within 2%, with the experimental results for a/W up to 0.6, as noted earlier. For deeper cracks, the two types of K/δ results diverge. Considering that independent analyses agree well for all crack lengths and experimental methods can be subject to unavoidable errors for deep cracks [*4*], the Eq 1 relation from analysis was used for the tests here.

Wedge-Load K_{Ic} *Tests*

Static load fracture toughness tests were performed at −60°C in T-L and L-T orientations using the configuration shown in Fig. 2, with minor modifications. Holes of 15-mm diameter

TABLE 3—*Crack-arrest fracture toughness test conditions.*

		Initial Conditions			Arrest Conditions		
Specimen/ Orientation/ Temperature	Side Groove, B_N/B	Notch: $(a/W)_0$	δ: δ_0, mm	K: K_0, MPa m$^{1/2}$	Notch: $(a/W)_a$	δ: δ_a, mm	K: K_{Ia}, MPa m$^{1/2}$
			AT −40°C				
5 T-L	0.76	0.34	0.97	197	1.00	...	...
			AT −60°C				
4 T-L	0.76	0.35	0.75	149	1.00	...	...
1 L-T	0.76	0.35	0.75	149	0.96	...	...
2 L-T	0.76	0.44	0.91	148	0.97	...	...
6 L-T	0.76	0.30	0.52	116	0.78	0.57	48
3 T-L	0.88	0.31	0.58	117	0.76	0.63	52
8 T-L	0.88	0.18	0.48	141	0.79	0.56	43
9 T-L	0.88	0.18	0.52	151	0.65	0.58	60
14 L-T	0.88	0.35	0.61	113	0.80	0.64	47
16 L-T	0.88	0.32	0.47	94	0.64	0.51	54
17 T-L	1.00	0.19	0.37	99	0.45	0.41	57
19 T-L	1.00	0.18	0.34	93	0.42	0.37	55
7 L-T	1.00	0.16	0.38	112	0.41	0.41	62
18 L-T	1.00	0.18	0.31	86	0.47	0.37	49

L-T mean K_{Ia}: 52.0 MPa · m$^{1/2}$, 6.2 standard deviation
T-L mean K_{Ia}: 53.4 MPa · m$^{1/2}$, 6.5 standard deviation
Grand mean K_{Ia}: 52.7 MPa · m$^{1/2}$, 6.0 standard deviation

were added for pin loading in fatigue to precrack the samples. Wedge load was applied quasi-statically until failure, as in a standard K_{Ic} test. Because of the inherent high stiffness of the wedge-load arrangement, the load-displacement trace changed quite abruptly as crack growth began; the trace showed a sharp drop in a manner very similar to that of the K_{Ia} test of Specimen 6 in Fig. 4. This resulted in an unambiguous determination of the critical K value for initiation of crack growth. Equation 1 and the same general procedures as used for K_{Ia} determination were used for determining K_{Ic}. For these tests, the wedge-load compact specimen arrangement was quite suitable for measurement of static plane-strain fracture toughness.

Results and Discussion

K_{Ia} *and* K_{Ic} *Results*

Tabular results of the K_{Ia} and K_{Ic} tests are listed in Tables 3 and 4. The final notch lengths marked by heat tinting and the related values of crack-arrest fracture toughness are shown in Table 3. Note that for the tests with $B_N/B = 0.75$ and a_0/W of 0.3 or more, only one test, that with the smallest a_0/W and δ_0, resulted in a proper arrest. Because a_0/W could be controlled directly, it was varied intentionally in subsequent tests, along with the planned variation in B_N/B. As expected, both higher B_N/B and lower a_0/W favored arrest, although a small change in a_0/W had a surprisingly large effect on arrest. This observation prompted a predictive analysis, described in an upcoming section.

Plots of wedge load versus crack-mouth displacement, δ, for two K_{Ia} tests are shown in Fig. 4.

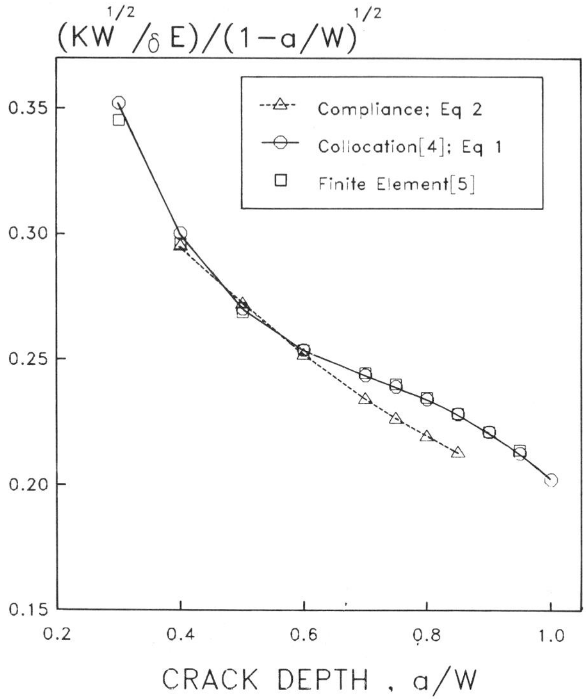

FIG. 3—*Comparison of* K/δ *results for wedge-load compact specimen.*

Specimen 6 had B_N/B and a_0/W as recommended by ASTM Test E 1221; Specimen 19 had a configuration outside the recommendations. As expected, Specimen 19 was much stiffer, but the general behavior and the resulting K_{Ia} (in Table 3) were quite similar. The overall K_{Ia} results, indicated by the individual and mean values in Table 3, show a relatively consistent crack-arrest toughness, with no readily apparent effects of material orientation, depth of side grooves, depth of initial notch, and initial applied K. The reason that the use of a much shallower initial notch than that recommended in ASTM Test E 1221 had no effect on the K_{Ia} results is believed to be the significant amount of crack growth that occurred beyond the initial notch.

The results of the static wedge-load fracture toughness tests are given in Table 4. Note that the results were not valid by the usual specimen thickness, B, criterion. A −60°C yield strength, σ_{-60}, of 757 MPa was used, which is 8% above the +20°C value from Table 1, based on results from the literature [6] for a similar steel. Even with this higher yield strength, the thickness criterion was not met. However, this is offset to some extent by the abrupt drop of load as crack growth began, noted earlier. The results of the static fracture toughness tests can be related to some features of the K_{Ia} tests. Note that the highest values of K_0 in the −60°C K_{Ia} tests, about 150 MPa · m$^{1/2}$, are about equal to the static toughness values. This is probably an indication that, for those tests, the crack grew through the brittle weld, stopped, and later reinitiated in the parent plate at $K \approx K_{Ic}$, to begin the run-arrest event. Some of the test traces showed a pop-in well before the point at which the crack ran, interpreted as a pop-in in the weld, which supports the above supposition.

Photos of fracture surfaces of three K_{Ia} specimens and one K_{Ic} specimen are shown in Fig. 5. Skewed crack growth was seen with four of the ten successful K_{Ia} tests. That shown for Specimen

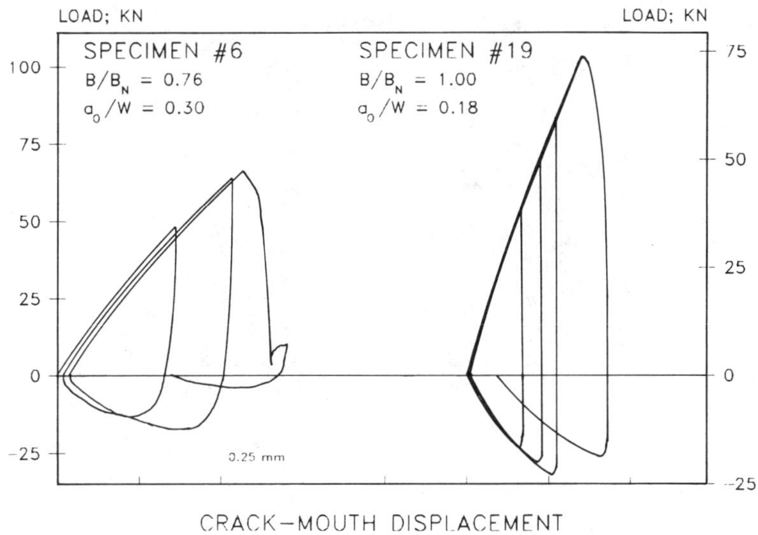

FIG. 4—*Wedge load versus crack-mouth displacement for* K_{Ia} *tests of BIS 690 at −60°C.*

6 was the worst observed; the slight degree of skew shown for Specimen 18 was typical. We believe the skewed crack growth was due to misalignment of the loading hole or the specimen support on individual K_{Ia} specimens, because it occurred sporadically for K_{Ia} tests and not at all for static tests.

A graphic summary of all the static and crack-arrest fracture toughness results is given in Fig. 6. The values of K_0 at the start of run-arrest, the K_{Ia} values, and the static test data are plotted versus their respective crack depths, a/W. The correspondence between the higher values of K_0 and the static fracture toughness data, noted earlier, can be seen. Regression analysis was performed to check for significant quantitative effects of test variables on K_{Ia}, including $(a/W)_a$, $(a/W)_0$, B_N/B, and K_0. Of these variables, only $(a/W)_a$ showed a correlation coefficient larger (in absolute value) than 0.5; its value was −0.66. Therefore, there is some decrease in K_{Ia} with increasing final crack depth, indicated by the solid line, from regression analysis. This decrease in K_{Ia} could also be attributed to an increasing amount of crack jump, because crack jump is not independent of final crack depth in these tests. These effects and explanations for a decrease in K_{Ia} have been noted before [*3*]. It is emphasized that had the K/δ relation from ASTM Test E 1221 been used to analyze these K_{Ia} results, the effect of decreasing K_{Ia} for deep cracks would

TABLE 4—*Fracture toughness from static wedge-load tests at −60°C.*

Specimen	Orientation	Fracture Toughness, $MPa \cdot m^{1/2}$	$[2.5 (K_{Ic}/\sigma_{-60})^2]/B$
a	L-T	147	1.88
b	L-T	111	1.08
c	T-L	142	1.76
d	T-L	156	2.12

Grand mean 139.0 $MPa \cdot m^{1/2}$; 19.5 standard deviation

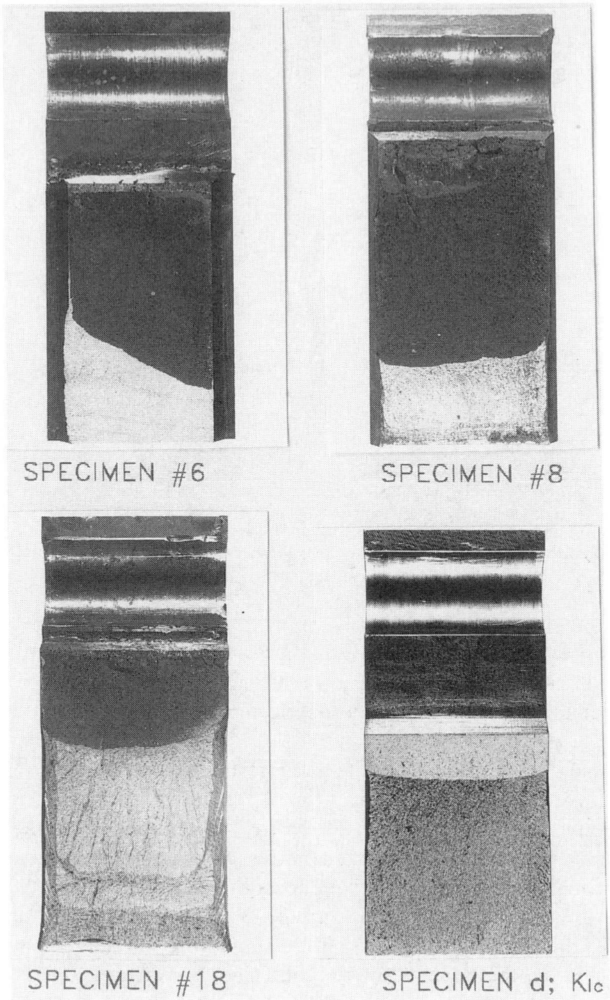

FIG. 5—*Photos of fracture surfaces of* K_{Ia} *and* K_{Ic} *specimens.*

have been more apparent. To demonstrate this point, the ten K_{Ia} results were recalculated using the ASTM Test E 1221 relation (Eq 2); the dashed line, a regression fit to these results, is shown, and an additional decrease of K_{Ia} with a/W can be seen.

Comparison with Other Results

The K_{Ia} results here were compared with those from other similar tests. Ripling and Crosley [*6*] tested AISI 1340 and 4140 steels at −54°C, a reasonably appropriate comparison to this work, although the yield strengths were somewhat higher in their work. In Table 5 some of their results are summarized. The results for 1340 steel, probably the more appropriate comparison, are in

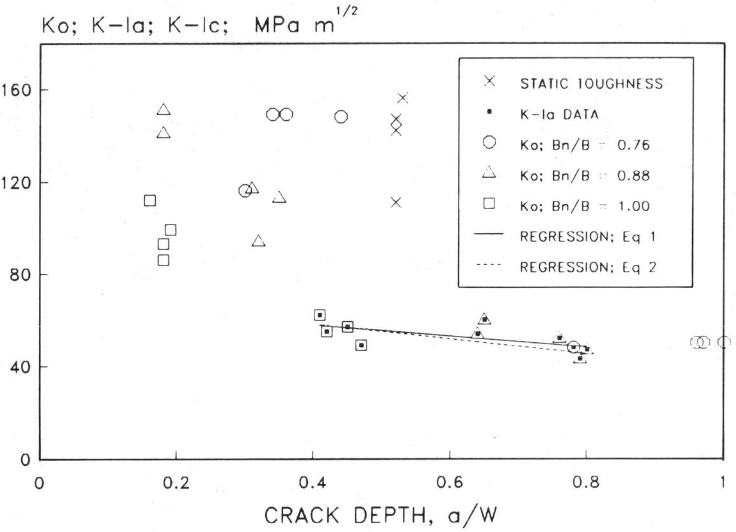

FIG. 6—K_0, K_{Ic}, *and* K_{Ia} *versus notch or crack depth.*

good agreement with the results here. Their 4140 steel results also agree well with the results here, except for their results from the lowest strength material. This could be explained by the significant transition with temperature that Ripling and Crosley noted in their K_{Ia} results.

An important difference between the overall results of this investigation and results of other types of rapid load fracture tests can be emphasized by comparison with results of dynamic *initiation* fracture toughness tests. Kendall [7] was among the first to investigate dynamic K_{Ic} in high-strength steels. He found no effect of loading rate in valid sized K_{Ic} results for AISI 4340 steel of 1275-MPa yield strength, tested at −51°C with a K rate of 10^5 MPa · m$^{1/2}$/s. Some recent work [8] compared static K_{Ic} with dynamic initiation K values determined from nonstandard J_{Ic} tests of 4340 vacuum-arc-remelt steel. They found typically *twofold increases* in dynamic initiation toughness compared to static when tested at K rates of 2×10^6 MPa · m$^{1/2}$/s over a wide range of temperature, from −140 to +100°C. Note that dynamic initiation toughness, K_{Id}, has been found [7,8] to be equal to or above K_{Ic} for this type of steel, whereas K_{Ia} is significantly below K_{Ic} in the tests here. This significant difference may be caused by the clear difference in fracture process, initiation of crack growth under rapid load in one case, and rapid run-arrest growth in the other.

TABLE 5—*Crack-arrest fracture toughness results of Ripling and Crosley* [6] *for two steels tested at −54°C.*

AISI 1340 Steel		AISI 4140 Steel	
σ_Y + 20°C, MPa	K_{Ia}, MPa · m$^{1/2}$	σ_Y +20°C, MPa	K_{Ia}, MPa · m$^{1/2}$
965	70	965	154
1100	51	1100	60
1240	50	1240	51

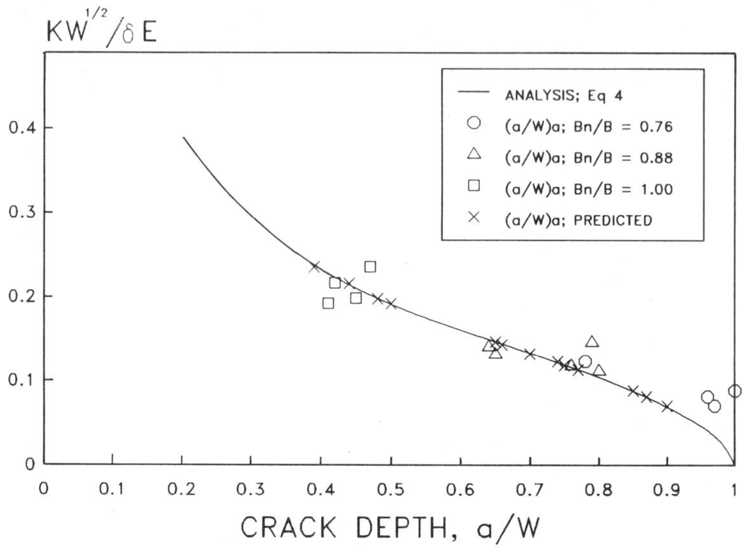

FIG. 7—*Calculated and measured values of* $KW_{1/2}/\delta E$ *versus crack depth.*

Prediction of Crack Depth at Arrest

The significant decrease in the arrest crack depth brought about by a small decrease in initial crack depth in these tests led to the following method of predicting the crack depth at arrest.

First, an expression for a/W in terms of the K parameter for the wedge-load compact, $KW^{1/2}/\delta E$, is required. This expression, essentially the inverse of Eq 1, was developed by regression analysis of data from Eq 1 and is given as follows:

$$a/W = f(V) = 1 + 1.132V - 47.29V^2 + 206.3V^3 - 359.2V^4 + 225.5V^5$$
$$\text{where } V = KW^{1/2}/\delta E \qquad \text{for } 0.15 \leq a/W \leq 1.00 \tag{3}$$

Equation 3 is compatible with Eq 1 within 0.02 W over the indicated range of a/W and within 0.007 W over the range $0.15 \leq a/W \leq 0.85$.

Using the expression of Eq 3, a prediction of crack depth at arrest, $(a/W)_a$, can be made as follows:

$$(a/W)_a = f(V_a)$$
$$V_a = K_{Ia}(B_N/B)^{1/2}W^{1/2}/(\delta_a/\delta_0)\delta_0 E \tag{4}$$

where the function, f, is from Eq 3. The effect of side grooving is accounted for by the $(B_N/B)^{1/2}$ term. Side grooves lower the specimens ability to arrest a crack, and this can be represented by an effective K_{Ia} equal to $K_{Ia} \times (B_N/B)^{1/2}$. The combination $(\delta_a/\delta_0)\delta_0$ represents the crack-mouth displacement at arrest, δ_a, obtained by using the experimental observation that δ_a is generally a bit larger than δ_0 by a constant ratio. For the tests here, the mean value of δ_a/δ_0 was 1.10 (see Table 2).

After-the-fact predictions of $(a/W)_a$ were made using Eq 4 to check the procedure. The results are shown in Fig. 7 for all 13 tests at −60°C, including the 3 in which the arrest crack depth was

TABLE 6—*Calculated arrest crack depth,* (a/W)a, *for various values of* $(a/W)_0$, K_{Ia}/K_0, B_N/B, *and for* $\delta_a/\delta_0 = 1.10$.

	$K_{Ia}/K_0 = 0.8$	0.7	0.6	0.5	0.4	0.3
		$B_N/B = 1.00$				
$(a/W)_0 = 0.2$	0.31	0.37	0.44	0.54	0.67	0.80
0.3	0.44	0.51	0.60	0.70	0.79	0.87
0.4	0.57	0.65	0.72	0.80	0.86	0.92
0.5	0.68	0.74	0.80	0.85	0.90	0.94
0.6	0.76	0.81	0.85	0.89	0.93	0.96
		$B_N/B = 0.76$				
$(a/W)_0 = 0.2$	0.37	0.44	0.52	0.63	0.74	0.84
0.3	0.52	0.60	0.68	0.76	0.84	0.90
0.4	0.65	0.72	0.78	0.84	0.89	0.94
0.5	0.75	0.80	0.84	0.89	0.92	0.95
0.6	0.81	0.85	0.89	0.92	0.94	0.97

beyond the $a/W = 0.85$ limit of ASTM Test E 1221. The open symbols are the values of $V_a = [K_{Ia\text{-}ave}(B_N/B)^{1/2}W^{1/2}/1.1\delta_0 E]$ plotted versus measured $(a/W)_a$, where $K_{Ia\text{-}ave}$ is 52.7 MPa $\cdot$ m$^{1/2}$ from Table 3; $(a/W)_a$, B_N/B, and δ_0 are from Table 3; and W and E are 0.100 m and 210 GPa, respectively. The predicted values, shown as an *X*, are the same values of V_a plotted versus the values of a/W calculated from Eq 3. The predicted values of a/W are in good agreement with the measured values for all but the deepest cracks. This is significant, because it indicates that the important effects of initial notch depth and side groove depth can be included in a mechanics-based prediction of the arrest crack depth before a K_{Ia} test is performed. However, because a measured value of δ_a/δ_0 was used, this prediction has its limitations.

Another, more general type of prediction of arrest crack depth can be made using the procedure outlined by Eq 4. By assuming various values of the ratio of K_{Ia} for the material of interest to the applied K at initiation of the run-arrest event, K_0, calculations of arrest crack depth can be made for various prescribed combinations of $(a/W)_0$, B_N/B, and δ_a/δ_0. Table 6 lists such calculations for the value of δ_a/δ_0 from these tests, 1.10, and two values of B_N/B. For tests in which the run-arrest begins from a crack in the parent material rather than at a brittle weld, $K_0 \approx K_{Ic}$. As noted earlier, this was the situation for some of the tests here. For $K_0 \approx K_{Ic}$, Table 6 can be used to make general predictions of arrest behavior for a given combination of material and test configuration. For example, for a material with K_{Ia} nearly equal to K_{Ic}, arrest is easy to manage even for a relatively deep initial notch and side grooves, as indicated by the first few columns in Table 5. For a material with K_{Ia} that is half or less of K_{Ic}, arrest is likely only for a relatively shallow initial notch and shallow or nonexistent side grooves, which were generally the configurations of the successful tests here.

Summary

The material characterization results of the investigation can be summarized as follows:

1. The grand mean crack-arrest fracture toughness, K_{Ia}, of ten tests of BIS 690 ship plate steel at $-60°C$ was found to be 52.7 MPa $\cdot$ m$^{1/2}$, with a standard deviation of 6.0 MPa $\cdot$ m$^{1/2}$. Individual mean values for T-L and L-T orientations were within about 1% of the grand mean, which shows no significant variation of K_{Ia} with orientation. The K_{Ia} results showed

no apparent effect of depth of side grooves or depth of initial notch. The results did indicate a slight decrease in K_{Ia} with increasing depth of crack at arrest. Regression analysis of the ten test results shows a decrease in K_{Ia} from 57.7 to 49.5 MPa $\cdot$ m$^{1/2}$ corresponding to an increase in $(a/W)_a$ from 0.41 to 0.79.

2. The mean static fracture toughness of BIS 690 steel at $-60°C$ determined from wedge-load tests similar in procedure and analysis to K_{Ia} tests was found to be 139 MPa $\cdot$ m$^{1/2}$. The test specimen thickness was equal to the full 50-mm thickness of the plate, but it did not meet the 2.5 $(K_{Ic}/\sigma_Y)^2$ validity requirement for a K_{Ic} test of this material at $-60°C$, calculated to be 84 mm.
3. The static fracture toughness of BIS 690 steel at $+20°C$ determined from J_{Ic} tests was found to be 239 MPa $\cdot$ m$^{1/2}$.
4. The K_{Ia} of BIS 690 at $-60°C$ was found to be a relatively small fraction of K_{Ic}, $K_{Ia}/K_{Ic} = 0.38$. This has implications for design and service life analysis of BIS 690 structural components subjected to low temperature. If service conditions allow a crack to run, an initiation fracture toughness approach to design and life analysis would be insufficient at best, possibly nonconservative.

The test method development results of the investigation are the following:

1. Crack-arrest tests with somewhat shallower initial notch depths than those recommended in ASTM Test E 1221, that is, in the range $0.15 \leq (a/W)_0 \leq 0.30$, were found to arrest at significantly shallower crack depths.
2. An expression for crack depth, a/W, in terms of $KW^{1/2}/\delta E$ and an associated analysis for predicting crack depth at arrest were found to give a good description for the BIS 690 test results at $-60°C$, including effects of side groove depth and initial notch depth on crack depth at arrest.
3. A static fracture toughness test procedure based on the wedge-load arrangement and analysis methods of ASTM Test E 1221 was found to be suitable for K_{Ic} tests of BIS 690 at $-60°C$. Aside from the common and unavoidable specimen thickness problems with this relatively tough material, the wedge-load K_{Ic} tests were consistent and easily interpreted.
4. Shallower initial notches and the expression and analysis for predicting crack depth at arrest are suggested as future additions to the ASTM Test E 1221 method for K_{Ia} tests. They address a persistent problem with the method, that being controlling and predicting the crack depth at arrest.

Acknowledgments

We are pleased to acknowledge M. Z. Shah Khan and N. J. Baldwin of Materials Research Laboratory for their help in fracture and mechanical testing and in developing wedge-loading test procedures, respectively.

References

[1] Hartbower, C. E. and Pellini, W. S., "Explosion Bulge Test Studies of the Deformation of Weldments," *Welding Journal Research Supplement*, Vol. 30, 1951, pp. 307s–318s.

[2] Davidson, T. E., Throop, J. F., and Underwood, J. H., "Failure of a 175mm Cannon and the Resolution of the Problem Using an Autofrettaged Design," in *Case Studies in Fracture Mechanics*, T. P. Rich and D. J. Cartwright, Eds., AMMRC MS 77-5, Army Materials and Mechanics Research Center, Watertown, MA, 1977.

[3] Rosenfield, A. R., Mincer, P. N., Marschall, C. W., and Markworth, A. J., "Recent Advances in

Crack-Arrest Technology," in *Fracture Mechanics: Fifteenth Symposium, ASTM STP 833*, R. J. Sanford, Ed., American Society for Testing and Materials, Philadelphia, 1984, pp. 149–164.

[*4*] Underwood, J. H. and Newman, J. C., Jr., "Comparison of Compliance Results for the Wedge-Loaded Compact Specimen," *Journal of Testing and Evaluation*, Vol. 16, No. 5, Sept. 1988, pp. 489–491.

[*5*] Pajot, J. J., to be published, General Electric Company.

[*6*] Ripling, E. J. and Crosley, P. B., "Crack Arrest Toughness of 4140, 1340 and 4340 Steel," MRL Report 792, Materials Research Laboratory, Inc., Glenwood, IL, Nov. 1981.

[*7*] Kendall, D. P., "The Effect of Loading Rate and Temperature on the Fracture Toughness of High Strength Steels," *Materials Research and Standards*, Vol. 10, No. 12, Dec. 1970.

[*8*] Chi, Y. C., Lee, S. H., Cho, K., and Duffy, J., "The Effects of Tempering and Test Temperature on the Dynamic Fracture Initiation Behavior of an AISI 4340 VAR Steel," Report 2, Army Contract DAALO3-88-K-0015, Brown University, Providence, RI, Aug. 1988.

DISCUSSION

E. J. Ripling[1] *(written discussion)*—In the study of the effect of side grooves on the measured values of crack-arrest fracture toughness (CAFT), the authors found that over a (B_N/B) range from unity to 0.76, the effect of side grooving was modest. This is certainly not always the case; this discusser found a much larger effect on tests of three 9% Ni steels as shown in Table D1.

This behavior difference in the two kinds of material suggests that it will be difficult, if not impossible, to find a relationship between CAFT values measured with and without side grooves. It also raises the question of when side grooves should be used in CAFT testing. This is discussed in Ref *D1* where the viewpoint is taken that full-thickness tests should be made whenever possible. This allows the least constrained metal near the smooth free surface to contribute to the measured toughness as it would in a structure fracture. If one is attempting to enforce a condition of plane strain, however, for example, in testing specimens that are thinner than the structure, preventing the contribution of the less constrained surface material is desirable, and side grooves should be used as done in ASTM Test E 1221.

TABLE D1—*Three 9% Ni steels tested.*

Plate Number	Plate Thickness, mm	Test Temperature, °C	Yield Strength, MPa	K_a, $B_N/B = 1$ MPa · m$^{1/2}$	K_a, $B_N/B = 0.75$ MPa · m$^{1/2}$
L25[a]	25	−162	943	189	119
L16	16	−196	906	430	220
		−162	816	>600	318
M28	28	−196	860	317	143

[a] L-T direction, all others T-L.

Reference

[*D1*] Crosley, P. B. and Ripling, E. J., "A Quality Control Test for Selecting Materials to Arrest Fast-Running, Full-Thickness Cracks," *Journal of Testing and Evaluation*, Vol. 18, No. 6, Nov. 1990, pp. 396–400.

[1] Materials Research Laboratory, Inc., Glenwood, Illinois.

TABLE D2—*Comparison of* K_a *for various values of* B_NB.

Material/Test	K_a, $B_N/B = 1$, $MPa \cdot m^{1/2}$	K_a, $B_N/B < 1$, $MPa \cdot m^{1/2}$	$(K_a/\sigma_{ys})^2/B$
	RIPLING DISCUSSION		
L25	189	119 ($B_N/B = 0.75$)	1.6
L16	430	220 ($B_N/B = 0.75$)	14.1
M28	317	143 ($B_N/B = 0.75$)	4.8
	UNDERWOOD ET AL.		
(averaged values)	56	48 ($B_N/B = 0.76$)	0.11
	56	51 ($B_N/B = 0.88$)	0.11

J. H. Underwood (author's closure)—Dr. Ripling makes important points in his discussion of the effects of side grooves on crack-arrest fracture toughness, in our results and his. Effects can vary from almost none, in our material and tests, to a factor of two or more decrease in crack-arrest fracture toughness as a result of 25% side grooves, as he showed. His advice on the use of full-thickness or side-grooved specimens was particularly noteworthy, essentially that the structural application dictates the choice.

Dr. Ripling's concern with finding a relationship between crack-arrest fracture toughness with and without side grooves may have at least a partial answer. A comparison of K_a for various values of B_N/B with the thickness criterion for K_{Ia} tests is revealing. See Table D2.

It is clear from Table D2 that our tests are well within the criteria for plane strain crack-arrest tests, whereas the Ripling tests are unavoidably outside the criteria because of the toughness level and thickness of the material. This could be the basis of a general expectation. For tests that meet the thickness criteria, side grooves may have relatively little effect on the result. Note that the L25 tests, which are least outside the thickness criteria, also show the least decrease in K_a caused by side grooves. This agrees with Dr. Ripling's arguments in that, as the dominant plain strain condition is approached, the contribution of the less constrained surface material becomes relatively less important.

Hugh J. MacGillivray[1] *and David F. Cannon*[2]

The Development of Standard Methods for Determining the Dynamic Fracture Toughness of Metallic Materials

REFERENCE: MacGillivray, H. J. and Cannon, D. F., **"The Development of Standard Methods for Determining the Dynamic Fracture Toughness of Metallic Materials,"** *Rapid Load Fracture Testing, ASTM STP 1130,* Ravinder Chona and William R. Corwin, Eds., American Society for Testing and Materials, Philadelphia, 1992, pp. 161–179.

ABSTRACT: A simple, reliable, and repeatable method has been developed for determining dynamic plane-strain fracture toughness using an on-specimen strain gage. This extends the range of the British Standard Method for Determination of the Dynamic Fracture Toughness of Metallic Materials (BS 6729) to include measurement at impact testing rates. Dynamic finite element computing was used to optimize the position of the gage on the specimen. In tests at low loading rates on Charpy specimens, good agreement was found between toughness values calculated from the applied load and the calibrated gage, and with those measured by the shadow optic caustic technique. At higher rates, several gage types and two adhesives were compared. A satisfactory method was developed for performing tests at high and low temperatures. The technique was applied to larger bend specimens of two steels, and a test program was conducted at seven laboratories to assess the influence of testing machine and instrumentation variables. From the results, a standard calibration and test procedure has been developed.

KEY WORDS: dynamic fracture toughness, fracture testing standards, impact testing, dynamic testing, fracture (metals)

Reliable standard dynamic toughness data are essential for evaluating the integrity of a wide range of components and structures. High loading rates generated by impact forces may occur only accidentally, but many military and transportation applications regularly experience loading rates many orders of magnitude greater than the quasi-static conditions under which most testing is conducted.

The primary objective of any proposed standard method for determining dynamic fracture toughness should be to provide accurate and reproducible results at loading rates that equal or exceed those encountered in service or projected emergency conditions. If the standard is to achieve general use, it must be the simplest method available, use existing testing machines and auxiliary equipment as far as possible, and use standard or near-standard test specimens to simplify planning and carrying out test programs. Because the tests may be performed only occasionally, the technique chosen should not need specially trained personnel or expensive specialized equipment and should not cause any unusual hazard. These requirements rule out a number of otherwise attractive techniques.

[1]Senior research officer, Imperial College of Science, Technology and Medicine, Mechanical Engineering Department, Exhibition Rd., London, SW7 2AZ England.

[2]Team leader, Structural Integrity, British Railways Board, Research Department, Brunel House, London Rd., Derby, DE2 8UP England.

TABLE 1—*Composition of steels used in UKWP dynamic fracture test programs.*

Composition, %	C	Si	Mn	S	P
BS 11 cast 3844	0.44	0.10	1.08	0.034	0.035
BS 4360-50D	0.18	0.30	1.50	0.040	0.040

The UK Working Party on Dynamic Fracture

The U.K. Working Party on Dynamic Fracture (UKWP) was formed in 1972 as the briefing group to the U.K. delegates of the International Institute of Welding Commission X. Since then, the UKWP has developed as an active, self-financed group of industrial, research institute, and academic workers. It acts as a forum for the exchange of information, initiates and conducts test programs, publishes results, drafts test standards, and serves as liaison with the British Standards Institution (BSI) and government departments. The UKWP has recently been reconstituted as part of the European Structural Integrity Society.

Its main achievements have been conducting the collaborative program on dynamic fracture, which led to the drafting, acceptance, and final publication by the BSI of the British Standard Method for Determination of the Dynamic Fracture Toughness of Metallic Materials (BS 6729). This was followed by a round robin on dynamic tension testing, which resulted in a draft test method.

Since 1985, the UKWP has been investigating methods for fracture toughness measurement at impact testing rates. After considering the options available, a second collaborative program was conducted using bend specimens with on-specimen strain gages in various locations. The results showed reasonable repeatability for both elastic and elastic-plastic behavior. It has been followed by a comprehensive test program on calibrated crack-tip gages supported by the British Department of Trade and Industry.

The First UK Collaborative Test Program

After the publication of some preliminary proposals [*1*], the UKWP conducted a program of 140 tests among 11 participating laboratories between 1976 and 1980. The tests were made at $\dot{K}$ rates between 10^3 and 10^4 MN $m^{-1.5}s^{-1}$, when the transducers were approaching their limits of response. Temperatures of -20 and $+100$°C were used to obtain elastic and elastic-plastic fracture behavior. The material used was a normal quality BS 11 rail steel whose composition is given in Table 1, and the specimen geometry was standard three-point bend with $W = 2B = 40$ mm. Servohydraulic machines were used with conventional load cells and clip gages. Force-time and clip gage-time signals were recorded independently, and an envelope procedure was applied to check linearity criteria. The effects of cross-plotting the two signals to obtain force-crack tip opening displacement (CTOD) records, and of external filtering was examined. A method of calculating CTOD that requires only the load-time and load-point displacement was developed and validated for BS 11 and two other steels. Detailed results were published [*2,3*] and used as the basis for writing BS 6729.

The Dynamic Testing Standard BS 6729:1987

This standard covers the determination of CTOD and K_{Ic} for metals at $\dot{K}$ rates between 2.5 and 3.10^3 MN $m^{-1.5}s^{-1}$. The Appendix A method without a clip gage extends the upper rate to 10^4 MN $m^{-1.5}s^{-1}$. These rates correspond to crosshead speeds in the range 0.02 to 100 mm/s. The provisions were described in some detail [*4*]. Although from inception a standard for dynamic testing, BS 6729 was designed to be broadly compatible with the existing British quasi-static fracture standards, British Standard Methods of Test for Plane Strain Fracture Toughness (K_{Ic}) of Metallic Materials (BS 5447) for K_{Ic} and British Standard Methods for Crack Opening Displacement (COD) Testing (BS 5762) for CTOD. There are many features in common with the ASTM Test for Plane-Strain Fracture Toughness of Metallic Materials (E 399) Annex A7 on rapid load testing.

Specimens may be of bend or compact tensile geometry for K_{Ic} determination, but bend only for CTOD, of maximum possible thickness to favor a valid result. Displacement control of the testing machine is specified, and requirements for the clip gage, roller contact bend, and tensile fixtures are given. The dynamic response of the load cell, clip gage, and recording equipment are specified, and no attenuation or filtering of the signals is allowable.

Requirement for Higher Rate Test Standards

The report by Baker [*5*] for the European Communities (EC) Bureau of Reference identified the need for a recognized method for determining fracture toughness under impact loading and discussed some of the techniques available. The cost of methods requiring high-speed photography was recognized as a disadvantage, and the need for a less expensive method was envisaged. Sumpter et al. [*6*] made estimates of the *K* rates experienced by ships and reported that, for the loading conditions of slamming, collision, and shock, these rates are greater than those covered by BS 6729.

Strain Gages in Dynamic Fracture Testing

The principal measurement methods available for dynamic fracture testing were reviewed by MacGillivray et al. [*7*]. Strain gages of both foil and semiconductor type have many advantages and are widely used for force measurement. Foil gages offer a large choice of geometry, low cost, easy application, stability, and reliability. An amplifier is normally required. Semiconductor gages are usually small with low mass and can be used unmounted so there is no backing to affect strain transmission. Because of their large gage factor, no amplifier is needed. Disadvantages include significant zero-drift with varying temperature. Strain gages can be mounted on the striker of the impact machine to measure the dynamic force applied to the specimen [*8,9*]. Unless the velocity is limited, as in the "low-blow test," some form of inertia correction [*9*] or fitting procedure [*10,11*] is required to interpret the oscillating signal and identify the crack initiation point.

Using the signal from a calibrated strain gage on the specimen to measure the applied load, overcoming the dynamic and inertia problems associated with striker load cells has been widely investigated. The review by Ireland [*12*] considered all the possible strain gage locations. Some are unsatisfactory because of dynamic bending effects in the specimen. The crack-tip gage location has advantages where fracture is mainly elastic because the local elastic strain increases almost linearly with increasing *K*.

The response of bonded gages to sweeping strain pulses was investigated by Oi [*13*] who generated a steep wave by the cleavage fracture of a notched tensile bar. For strain amplitudes of about 1000 microstrain, he measured rise times of less than 2 μs using wire strain gages on polyester backing. An expression for estimating the rise time was derived related to the gage length and the speed of sound in the material. Brickle [*14*] developed Oi's work by using gages of different lengths mounted on an aluminum bar and loaded by a transient strain pulse generated from an explosive detonator. A correction technique was derived to compensate for the effect of gage length. Some details of the gages were given, but the adhesive and bonding technique was not stated.

Loss et al. [*15*] described dropweight tests at a velocity of 1.7 m/s on bend specimens of A212-B steel. The specimens were instrumented with two 3-mm gages in positions close to the crack tip and calibrated statically. In one case, the striker load was also measured during the impact. The results for a range of test temperatures between −84 and −12°C indicated time to fracture (t_f) in the range 120 to 400 μs. K_{ID} was calculated from the point of sudden load drop using the static calibration, and differences were reported from the two gage positions ascribed to plasticity. The load trace showed no sign of the fracture initiation. Suggestions were made for deliberately attenuating the striker signal and imposing a minimum allowable value on t_f.

Giovanola [*16*] suggests using gages smaller than 2 mm for measuring fracture properties in impact one-point bending and placing them close to the crack tip but outside the plastic zone. Mines and Dutton [*17*] successfully used a small crack-tip gage for comparison with inertial modeling of Charpy specimens in a Hopkinson pressure bar. Mall et al. [*18*] also used crack-tip gages on large steel and aluminum bend specimens and compared results with an elastodynamic finite element (FE) program. Good agreement was found for brittle and ductile fractures in steel, but variations occurred in aluminum, which were ascribed to extensive plasticity.

Joyce and Hackett [*19*] and Curry et al. [*20*] placed gages on the back and front faces of large bend specimens and encountered some problems with oscillation at impact velocities up to 4 m/s. Recent work by Couque et al. [*21*] on initiation toughness of reactor steels used a pair of compact tension-type specimens instrumented with near crack-tip gages, in a Hopkinson bar apparatus, as well as crack propagation gages and eddy current displacement transducers to measure CTOD directly. The crack-tip gages were not calibrated, reliance being placed on the Hopkinson bar. Excellent results were reported at K rates up to 4.2×10^6 MN $m^{-1.5}s^{-1}$. Brun-inghaus et al. [*22*] placed their calibrated gages on the arms of the compact specimen and found little problem with inertial effects.

Dally and Sanford [*23*] describe a method using one or more strain gages that avoids the need for static calibration. Multi-parameter solutions relating K_I and the near-crack tip strain field have been developed in terms of the Westergaard stress functions, and the gage positions and angular alignment required to minimize measurement errors have been determined. Experimental verifications have achieved better than 5% agreement between K_I calculated from strain gages and applied load.

Sharpe et al. [*24*] used laser interferometry to measure dynamic CTOD. The authors state how much more useful their results would be if they had a direct measurement of dynamic load, rather than making assumptions about its linear increase with time.

Strain gages have been used in conjunction with shadow optic caustics. The technique, developed for dynamic fracture by Kalthoff [*25*] and Hermann [*26*], reflects a beam of light from the concave depression formed at the crack tip as the specimen is loaded. This produces a shadow image, which can be photographed by a high-speed camera. The size of the caustic is related directly to the crack-tip stress intensity factor. The technique has been used to develop impact response curves for linear elastic behavior, characteristic of any material/geometry combination. Then only a measurement of t_f is required, using a crack-tip gage or magnetic pickup, to determine K_{ID}. The main disadvantage is the need for specialized optical and high-speed camera equipment.

The Second U.K. Collaborative Test Program

Having identified the on-specimen strain gage technique as the possible basis for a higher rate test standard, the UKWP in 1984/1985 conducted a feasibility study. Dynamic tests were made by six laboratories on specimens manufactured by The Welding Institute. The details of the two steels used, BS 11 and BS 4360-50D, are given in Table 1. All tests were conducted at room temperature on Charpy and bend specimens with gages bonded in various locations, at impact velocities up to 5 m/s. The results were encouraging and used in planning the detailed test program described below.

The Test Program on Crack-Tip Strain Gages

This project was designed to investigate in detail the technique of using a calibrated strain gage bonded to the specimen to determine the dynamic fracture toughness of metallic materials. The objective was to extend the provisions of BS 6729 to the highest impact loading rates available using fast servohydraulic, dropweight, and pendulum testing machines. Full details are included in the final report [*27*]. To achieve the maximum participation and use available test specimens, the program was limited to testing bend geometry using the same two steels.

The strain gage is required to perform several functions. In fracture toughness tests, it provides a force-time record that allows determination of maximum force, show the limits of force variation, a measure of K rate, and an indication of arrested crack extension or pop-in. The gage can also give the force and time/displacement data required for calculating CTOD by the BS 6729 Appendix A method.

Dynamic Finite Element Analysis

The two-dimensional elastodynamic FE program *Bolknak* developed by Keegstra [*28*] and progressively modified by Curr et al. [*11*] was used. Contact stiffnesses were determined by fitting the computed striker force record with analytical models and experimental data.

Optimizing the Strain Gage Position

To optimize the gage positions and predict expected strains for a range of K_c values, a mesh was developed using constant strain triangular elements with mesh size increasing with distance from the crack plane as shown in Fig. 1.

The strains in a Charpy specimen in the longitudinal (y) and transverse (x) directions were investigated. These results, plotted in Fig. 2 for a striker velocity of 5 m/s, indicated that a gage positioned in the y orientation at a point level with the crack tip and 5 mm away would experience minimum oscillation and reasonable strain levels. The back face gage position gave large oscillations in both orientations.

A similar procedure was used to model impact on the large three-point bend specimen with $W = 2B = 40$ mm, span of 4 W and an overall length of 200 mm. The plot K against time for a velocity 5 m/s is shown in Fig. 3, which is thus an *impact response curve.* The curve rises smoothly after a delay of about 5 μs from $t = 0$ at striker impact. After 100 μs and at a K of 97 $MNm^{-1.5}$, the first node was released and the K dropped sharply, then in steps as successive nodes were released.

The longitudinal strain 10 mm from and level with the crack tip is also plotted. This increased simultaneously with the crack-tip K and showed only slightly more oscillation up to the point of sudden fall, which occurred about 2 μs later than first node release. After the fracture, the strain trace became compressive, then rose to zero, as observed in many experimental records. The predicted striker force was of the typical inertia peak form in which the maximum value depends on striker velocity, mass, and contact stiffness.

Developing the Experimental Technique

A series of tests was made to determine the effect of strain gage and adhesive type on the results of dynamic tests and to assess the dynamic response, the effect of test temperature on

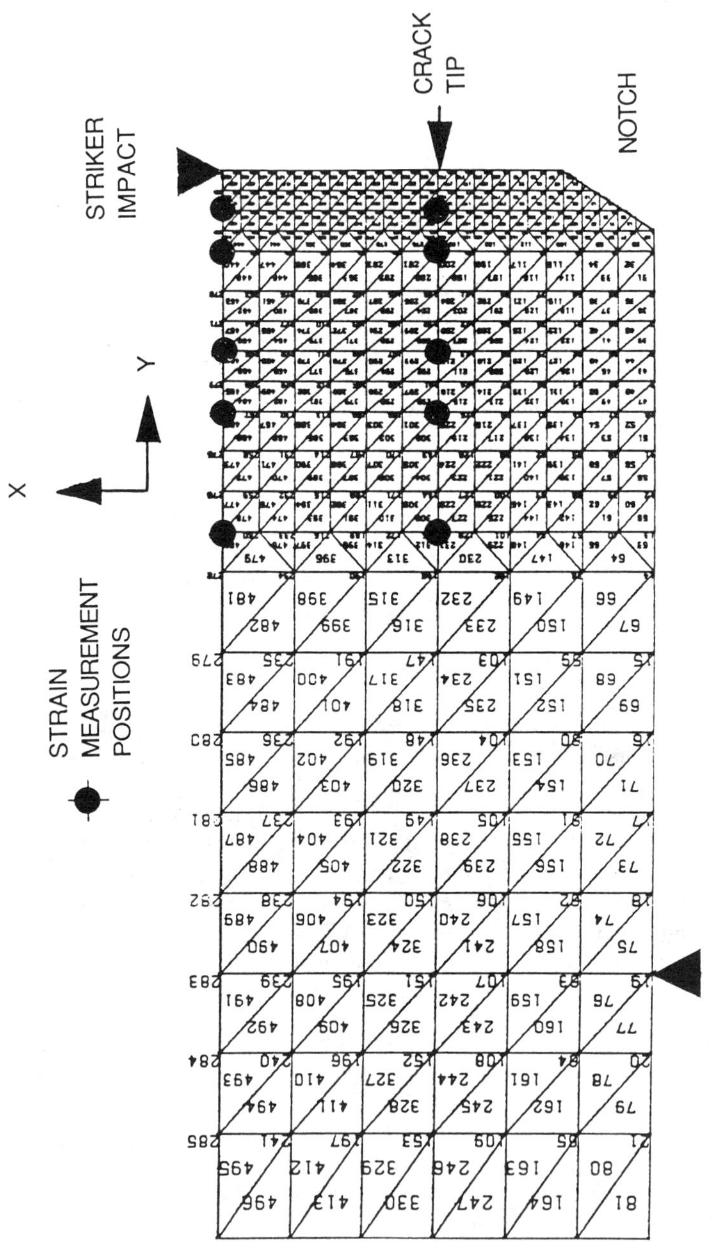

FIG. 1—*Typical half-Charpy specimen mesh for elastodynamic finite element program* Bolknak *showing location of strain measurement positions.*

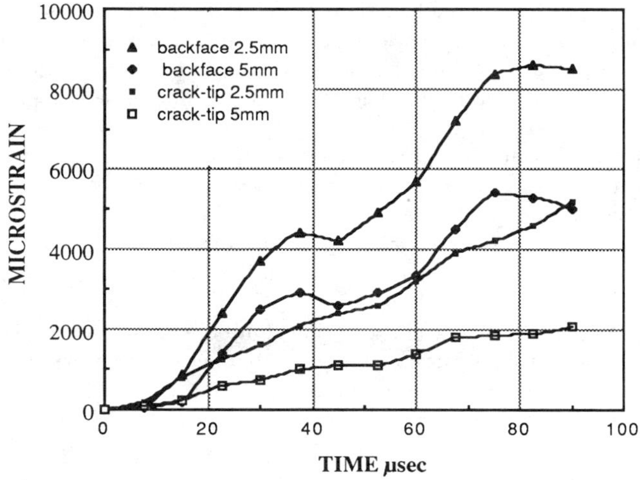

FIG. 2—*Comparison of computed strains in* y *direction at positions 2.5 and 5 mm from crack on back face and in line with crack tip for 5-ms*$^{-1}$ *Charpy impact.*

response and calibration, and general ease of application and use. To economize on material, initial testing was confined to Charpy specimens of both steels.

Strain Gage and Adhesive Selection

Four gage types were chosen as representing the widest variation of properties and performance. These were constantan/polyimide, constantan/epoxy, and constantan/phenolic for tem-

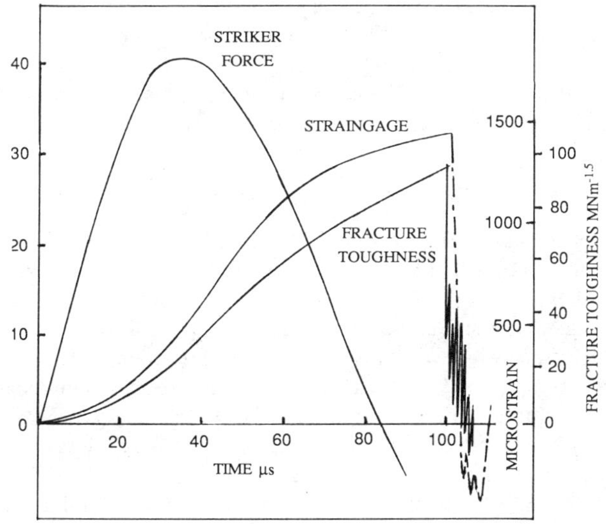

FIG. 3—*Computed impact response curve of dynamic fracture toughness, crack-tip strain gage response, and striker force for 5-ms*$^{-1}$ *impact on bend specimen.*

peratures from −200 to +200°C and K-alloy/phenolic with a range of −270 to +400°C. A gage grid size of 1.5 mm was chosen as a reasonable compromise between ease of handling and the small physical size required for point strain measurement for comparison with computed results and for use on small Charpy specimens. The grid shape for all gages was approximately square, and gage resistance was 350 Ω.

Two adhesives were selected having properties at opposite ends of the range of performance and ease of use. The first was a catalyzed methyl-2-cyanoacrylate (cyano) chosen for its fast room-temperature curing and recommended for static and fatigue use between −185 and 95°C. The second adhesive was a solvent-thinned two-part epoxy that required a high-temperature cure under clamping pressure. It gave a thin, even glue line and had a temperature range of −269 to 230°C.

All specimens were fatigue precracked as specified in BS 6729 to give a fatigue crack a/W in the range 0.45/0.55. Surface preparation and gage bonding followed the manufacturer's standard recommendation. Gages were placed level with the crack tip with the center of the grid 5 mm from the crack.

Instrumentation and Calibration

The striker and instrumentation for the Charpy tests were as described previously [*7,8*]. The dynamic response of the amplifier-lead-recorder system was checked by injecting a sine-wave signal. A reduction of 1 dB was found at 500 kHz, which was satisfactory because the highest frequency load oscillations for 5 m/s impact were at about 50 kHz.

The strain gage amplifiers used were Fylde Electronics FE359TA configured for three-wire quarter bridge operation. Gain was 100 and bridge voltage was 5 V, calculated to give optimum output without thermal drift. Dynamic response of the gage-amplifier system was found to be −3 dB at 300 kHz. Static calibration of the strain-gaged specimens, accurately centered on the anvils of the Charpy machine by means of a locating tool, was performed in situ up to the maximum final precrack load by a proving ring and stirrup. The calibration was very sensitive to specimen location. For testing, the specimens were clamped by light springs against the anvils to prevent the lead pulling the specimen out of position.

One static test to fracture was performed. The load was applied manually and measured directly. From the measured force, the stress intensity factor was calculated using the conventional quasi-static formulation in BS 6729. Fracture occurred at a K_c of 47.2 MN $m^{-1.5}$, whereas using a least-squares fit to the strain gage calibration, extrapolated to give a equivalent force, gave a K_c of 44.2 MN $m^{-1.5}$, an under-reading error from the strain gage of 6.5%.

Strain Gage and Adhesive Performance

A series of Charpy tests was conducted at 5.3 m/s on BS 11 specimens. Each carried two gages, one constantan/polyimide and the second one of the other types for comparison. Four specimens used cyano adhesive and the other four epoxy. A typical set of load- and strain-time records is shown in Fig. 4. The striker load shows an inertia peak with no indication of fracture, whereas the strain gage traces increase linearly and show a sudden drop at the cleavage point. The results are shown as a plot of K_c against t_f in Fig. 5. For the gages bonded with cyano, there was a high scatter in K_c and an unexpected variation in t_f. All gage types showed much closer grouping of K_c and t_f when bonded with epoxy. The predicted strain levels at fracture of about 700 microstrain were achieved.

This apparent strain rate sensitivity of cyano may be due to modulus variation with glue thickness. Because some thickness variation is unavoidable with the bonding method used, epoxy must be the preferred adhesive and was used for future tests.

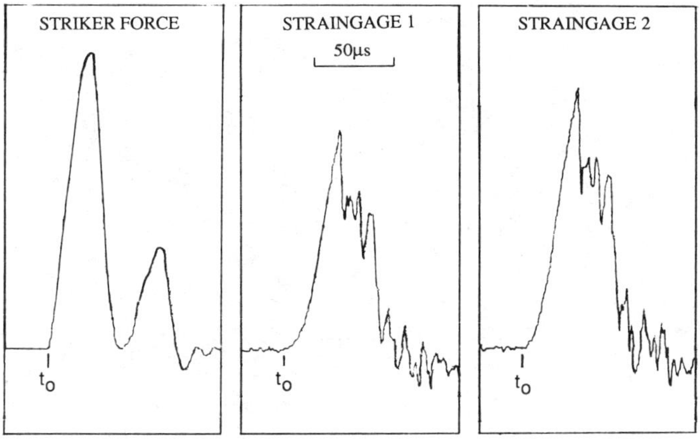

FIG. 4—*Striker force and crack-tip strain gage records for Charpy test at 5.3 ms^{-1}.*

The four gage types showed no identifiable variation in performance, and because the constantan/polyimide gage with large copper tabs is easy to use in small sizes, it is the preferred type, with the K-alloy/phenolic gage available for extreme temperatures.

Low-Blow Tests on BS 11 Charpy Specimens

Low-blow Charpy tests were made on two BS 11 specimens with one or two crack-tip gages at an impact velocity of 1 m/s, when the testpiece did not transmit severe inertial oscillations to the striker load cell, and the load record could be used for comparison. Agreement within ±2 MN m$^{-1.5}$ was found between the K_c values calculated from load and strain gage records at a K rate of 2×10^5 MN m$^{-1.5}$s^{-1}.

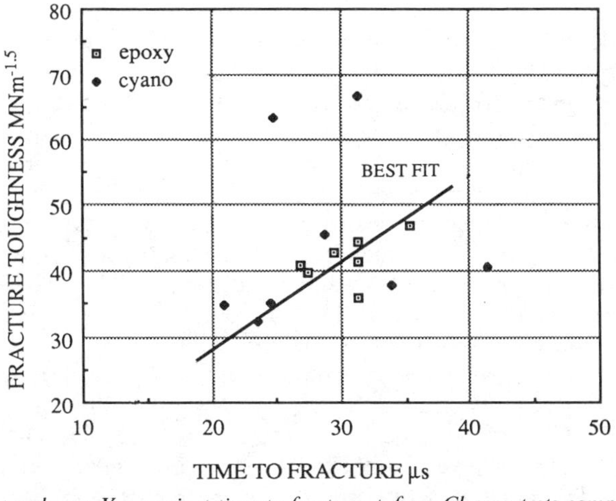

FIG. 5—*Fracture toughness,* K_c, *against time to fracture,* t_f *from Charpy tests comparing scatter of results from epoxy and cyano-bonded strain gages.*

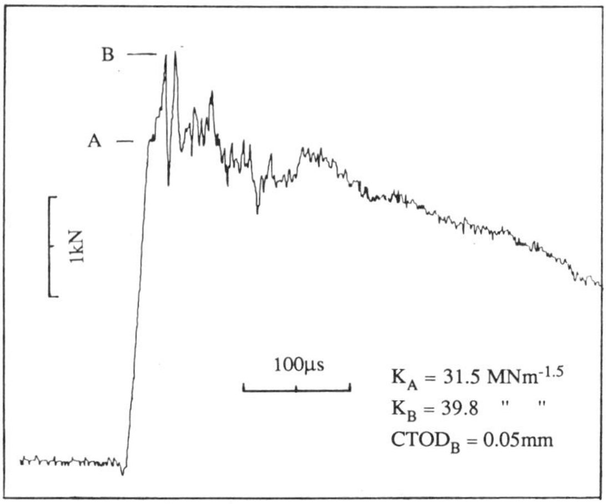

FIG. 6—*Crack-tip strain gage response for 5.3-ms^{-1} impact at −25°C on BS 4360-50D Charpy specimen.*

Shadow Optic Caustic Tests

A number of tests were performed in a servohydraulic machine in displacement control, at rates up to 0.1 m/s, in which simultaneous measurements of stress intensity were obtained from strain gages and by the shadow optic caustic method. Charpy specimens of BS 11 were polished on the side opposite the strain gages, illuminated by a laser and the caustic image filmed at up to 5000 frames per second. The results showed considerable scatter in K_c between different specimens but agreement within ±1 MN m$^{-1.5}$ between the caustics and crack-tip gages.

Low-Temperature Charpy Tests on BS 4360-50D

These had the aims of investigating the experimental problems of using gages at low temperatures and analyzing the gage response through the temperature transition zone. The BS 4360-50D specimens were instrumented with one gage and calibrated. Problems were encountered when cooling the specimens using a solvent bath cooled by liquid nitrogen. The solvent affected the gage bond strength, and standard coatings were unable to prevent failure. However, even uncoated gages were not affected by liquid nitrogen alone. A test technique was developed involving cooling the specimen, with a thermocouple welded close to the fatigue crack tip, in a liquid nitrogen flask to −196°C. It was then positioned in the Charpy anvils, the thermocouple meter was monitored, and the pendulum was released at the required temperature, which could be judged within ±1°C. The rate of temperature rise was such that tests could be performed as low as −100°C. A check made at −40°C established that there was no measurable change in gage calibration due to gage factor variation.

The low temperatures hardened the leadwire insulation, which caused some failures of the solder connections. This problem was reduced by attaching the lead at an angle to minimize bending. The thermocouple lead acted as an aerial, inducing electrical interference of up to 200 mV. The problem was overcome by spot-welding a ground lead close to the strain gage and by switching off the thermocouple meter before releasing the pendulum.

Tests were conducted over the temperature range −50 to 0°C at 5.3-m/s impact velocity. Both the test records and fracture appearance showed increasing ductility with increasing temperature, but in all cases the fracture mode was cleavage. Analysis used both K and CTOD approaches. The CTOD results showed no trend with temperature, giving values around 0.07 mm as expected where only limited plasticity occurred before fracture.

Figure 6 shows the crack-tip gage trace for a test at −25°C. There is a distinct change of slope

TABLE 2—*Details of the high-rate testing machines and instrumentation used in the UKWP interlaboratory test program.*

Imperial College	Instrumented dropweight m/c Instrumented Charpy 300-kHz Fylde amplifiers + 5-MHz Datalab recorders	to 6 m/s to 5.3 m/s	6 kJ 300 J
The Welding Institute	Drop tower Standard Charpy 170-kHz AD524 amplifiers + 2-MHz Datalab recorder	to 13 m/s 5 m/s	4.5 kJ 300 J
British Steel	Instrumented dropweight m/c Standard Charpy 90-kHz Fylde amplifiers + 2-MHz Datalab recorder Hewlett Packard computer data analysis system	to 7.1 m/s 5.5 m/s	1.0 kJ 294 J
Ruhr University	Wolpert instrumented Charpy 1-MHz amp and 300-kHz Fylde amp + 2-MHz Nicolet + computer	to 5.5 m/s	300 J
British Gas	Large pendulum m/c Standard Charpy 50-kHz Fylde amplifiers + 20-MHz Gould recorder	to 6.4 m/s 5.25 m/s	8 kJ 300 J
CEGB-CERL	ESH servohydraulic m/c 300-kHz Fylde amplifier + 2-MHz Datalab recorders	to 5 m/s	250 kN
British Rail	Large drop tower Standard Charpy 30-kHz CML amplifiers + 1-MHz SE Labs tape recorder 100-MHz Tektronix storage oscilloscope	to 15 m/s 5 m/s	90 kJ 300 J

at Point A, and the signal then increased to Point B where fracture occurred. The value of K at points was calculated, also the elastic and total CTOD to Point B according to the procedure detailed in BS 6729 Appendix A. After the initial cleavage at Point B, the gage output increased again and only decreased slowly in an apparently ductile manner. The form of the striker load trace, however, was similar to Fig. 4 and was typical of brittle behavior. Hence, the crack-tip strain gage is able to sense small changes in specimen behavior, whereas the striker load cell signal is dominated by inertial effects.

Trials on Large Bend Specimens

Trials were conducted at Imperial College on bend specimens of both steels to gain experience in gage calibration techniques and conduct of high-rate tests. Variations in gage position and loading rates gave an opportunity to confirm the dynamic FE predictions. The equipment used is listed in Table 2.

Interlaboratory Test Program

All the specimens were prepared at Imperial College using BS 11 and BS 4360-50D steel remaining from the earlier UKWP test programs. Three-point bend specimen sizes were 20 by 40 by 200 mm for BS 11, and 18 by 36 by 200 mm approximately for the 50D. After fatigue precracking, 1.5-mm constantan/polyimide gages were bonded using epoxy adhesive as previously described and positioned with their centers 5 or 10 mm from the crack tip for the Charpy specimens and 10 or 20 mm for the bend specimens.

The tests were carried out between May 1988 and January 1989 at the laboratories of The Welding Institute; British Rail Research, Derby; British Steel Technical, Rotherham; British Gas

TABLE 3—*Comparison of calibrations for Specimen B 102 at four testing laboratories.*

British Steel	...	0.0278 kN/mV	
Ruhr University	0°	0.0338	
	180°	0.0319	0.0329 av
British Gas	0°	0.0268	
	180°	0.0277	0.0273 av
British Rail	...	0.0313	
Arithmetic mean (6)		0.0299	Sample standard deviation 0.0028 or ± 9.4%

Engineering Research Station Killingworth; Central Electricity Research Laboratories (CERL), Leatherhead; and the Ruhr University, Bochum, Germany. Static calibrations and impact tests at up to 15 m/s were performed using the range of servohydraulic, dropweight, and pendulum impact machines detailed in Table 2. Test temperatures ranged from −50 to +50°C. The performance of instruments and recording systems was compared, and the practical problems involved in dynamic fracture testing were assessed. A total of 117 calibrations and 40 successful tests was performed. The failure rate was 20%.

Results

Specimen Calibration

Reliable and repeatable calibration is fundamental to the success of this test method, which is particularly sensitive to error because specimens can be loaded only to the maximum precracking force, which must not exceed 0.7 of the final K_{Ic}. The calibration slope is calculated by the least-squares method and extrapolated to find the force at fracture.

The calibration includes the effects of crack length and shape, gage location, specimen dimensions and squareness, bend rig span and alignment, and load cell accuracy. Variations must be expected between individual specimens and when one specimen is loaded in several machines. Multiple calibrations of Charpy specimen B102 are shown in Table 3, where the variation was less than ±10%. The results also show the small effect of performing a second calibration after turning the specimen through 180°.

The average calibrations for all specimens are given in Table 4 and show variations ranging from 9.5 to 23%. The larger variation for the small specimens is reasonable because errors in gage positioning and alignment are magnified.

In-situ calibration followed directly by testing without specimen removal is the ideal procedure. If this is not possible, care must be taken to ensure near-perfect alignment in both the

TABLE 4—*Average calibrations for all bend and Charpy specimens.*

Specimen	Gage Distance from Crack Tip, mm	Number of Specimens	Number of Calibrations	Mean Calibration, kN/mV	Standard Deviation, %
BS 11 bend	10	17	28	0.1018	9.5
50D bend	10	12	23	0.0834	13
50D bend	20	7	15	0.2272	12
BS 11 Charpy	5	16	35	0.0283	18
50D Charpy	5	2	6	0.0321	6.5
50D Charpy	10	2	10	0.1832	23

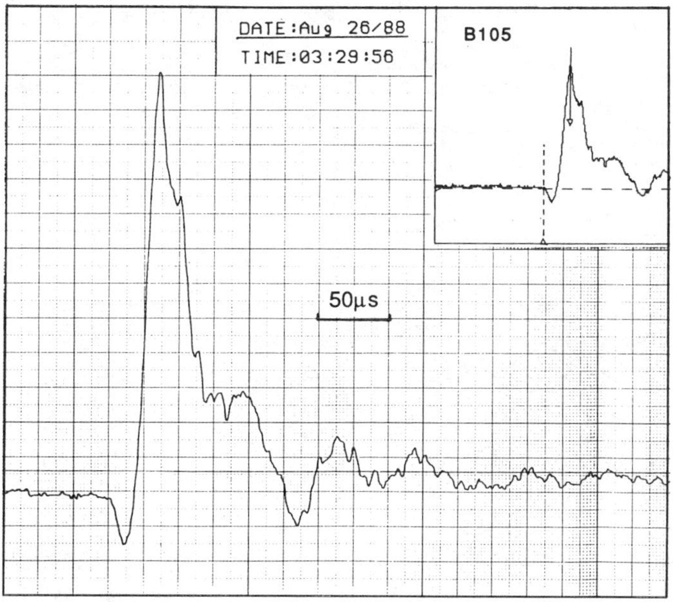

FIG. 7—*Crack-tip strain gage record for 5.3-ms^{-1} Charpy impact test showing initial dip in trace.*

calibration rig and the testing machine. Calibration in 0 and 180° positions provides a check on this, as does use of a standard calibration specimen.

Strain Gage Position

The initial tests on Charpy specimens used a gage with its center 5 mm from the crack tip. Because for a Charpy specimen W is 10 mm, this gage position is referred to as the $W/2$ position. It was anticipated that, for fully ductile fracture in Charpy specimens, the gage should be positioned at least 10 mm from the crack tip at the W position resulting in the reduction in sensitivity by a factor of six. This is shown in Table 4, where the calibrations are about 0.03 and 0.18 KN/mV, respectively. The average offsets after testing, indicating yielding under the gage, which were small for BS 11 at $W/2$, increased for 50D even at the W position.

For the BS 11 bend specimens, a gage position 10 mm from the tip, or $W/4$, was practical. Again, for 50D, this position was too close, but the 20-mm or $W/2$ position gave acceptable final offsets. If a single gage position must be specified, $W/2$ is most satisfactory, with $W/4$ as an alternative when negligible plasticity is anticipated.

Initial Dip in Strain Gage Record

For a number of the tests, an initial dip or compressive pulse of up to 100 microstrain was noted in the strain gage trace. This effect can be seen in Fig. 7. The dip causes uncertainty in both maximum load and t_f measurement. It is a dynamic effect, not observed during static calibration. Curry et al. [*20*], testing large bend samples fitted with ¼-span front-face gages at 4 m/s, recorded a dip of about 20% of the maximum load value. This was explained as probably due to a compressive stress wave from the loaded center portion of the specimen before the hinged-beam mode of deformation was established. Similar behavior was noted by Mall et al.

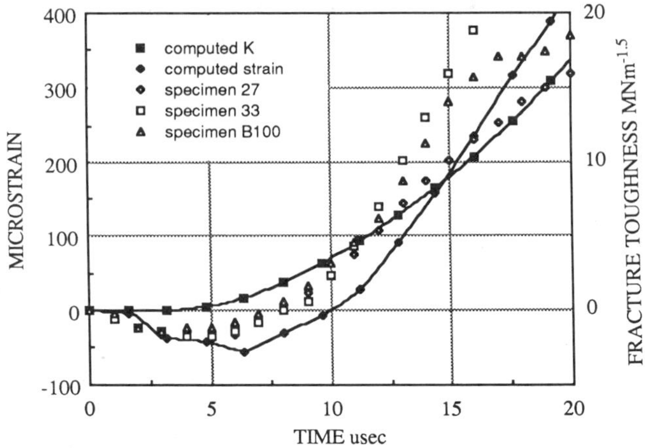

FIG. 8—*Comparison of computed and experimental results for offset impact.*

[*18*] for both experimental and computed dynamic strains at the ¼-span front-face location, but no dip was found for the gage location 8 mm from the crack tip at impact velocities between 2.5 and 8.6 m/s. An analysis of stress waves during impact over a subsurface crack by Brock et al. [*29*] showed that a negative Mode 1 stress intensity factor exists for a short time after impact, although the magnitude appears to be too small to cause the strain levels recorded.

The effect may be due to misalignment during the impact, causing the specimen to be struck off-center. The guides of dropweight machines allow side play, and lateral vibration of the striker is possible in a pendulum machine. The effect was not seen on the tests using a servohydraulic machine in which the actuator gave precise guidance.

To investigate the cause, additional dynamic FE computations were made. The complete Charpy specimen was modeled because the geometry is not symmetrical. The gage position was 5 mm from the crack tip, and a number of offset striker impact positions were modeled at 5 m/s. Figure 8 shows computed K_c and gage strain for an impact 2.5 mm from the crack tip, on the opposite side from the gage. The gage response was initially compressive, showing a dip to 50 microstrain. The K record was little affected. Experimental strain records from three BS 11 Charpy tests are plotted for comparison and show similar behavior. In the case of impact on the side of the crack closer to the gage position, the K response was little changed and the strain did not show a dip but increased initially more steeply up to 20 μs.

Thus, although the dynamic K response is not affected by the exact impact position, the initial output of the crack-tip gage is strongly influenced. This imposes limitations on measuring K at short t_f and on the use of the gage signal for determining K rate. If the initial dip cannot be prevented, a limiting value of some small percentage of maximum load should be imposed. The t_f taken from the beginning of the dip was found to give a more realistic value than that from the slope of the linear portion of the trace, which conflicts with the requirements of BS 6729 for calculating dK/dt.

Signal Noise

Severe problems with pick-up of continuous airborne electrical noise were experienced at some laboratories, caused by machinery operating in the same building as the testing machine. Precapture filtration of the signal is prohibited by BS 6729 to avoid loss of vital detail. One solution is to test only when the noise level is low. If this is not possible, noise reduction measures must be taken using amplifiers with high common-mode rejection and twisted or woven screened cable.

Instrumentation

All the instrumentation used in the test program and listed in Table 2 was satisfactory. The load cells used for specimen calibration were not all to British Standard Specification for the Grading of the Forces Applied by Materials Testing Machines (BS 1610:1985) which may account for some of the variation. The strain gage amplifiers also performed well. The fastest signal rise time encountered was about 25 μs for the BS 11 Charpy tests, requiring a 1-dB frequency response of only 10 kHz according to BS 6729 Appendix B3. All the amplifiers were considerably faster than this with 1-dB response from 40 to 500 kHz, but it is recognized that a higher performance than BS 6729 may be required for this type of test. If the Oi [*13*] results for response time of strain gages are correct, the limiting t_f for this test method is around 2 μs, although with very short rise times it will be necessary to allow for the delay in the shear wave arrival. Impact velocities around 5 m/s give t_f in the range of 40 to 60 μs for the large bend specimens. Scaling thus suggests a maximum impact velocity around 125 m/s and a K rate of 3.10^7 MN $m^{-1.5}s^{-1}$. To capture a reasonable number of data points during this rise time requires a recording rate of at least 20 MHz.

Recorder triggering was generally from the strain gage signal and caused fewer problems than expected. However, for regular use, optical or infrared triggering from the striker position is repeatable and immune to electrical interference.

Test Procedure

Strain gages were bonded in batches, each one taking about 15 min. Setting up the calibration and impact test rigs took one or two working days. Calibrating each specimen required about 30 min, and impact tests could be made at a rate of one per hour.

The computerized recording and analysis system at British Steel was easy to use and produced report-quality results like Fig. 9 within minutes. The minimum mistakes and delays were encountered when operating instruments normally used together. There is a strong case for any laboratory seriously involved in high-rate testing to assemble a set of amplifiers, recorders, and data storage equipment optimized for that purpose. A standard written test procedure and record sheet are needed to ensure that no steps are missed.

Dynamic Fracture Toughness Results

All the results from the program are given in Table 5. The mean values and deviation for BS 11 are similar for both bend and Charpy specimen geometry. All BS 11 results are plotted against K rate in Fig. 10, together with the original UKWP static and low-rate results. The new high-rate results continue the trend of decreasing K_{Ic} with increasing rate, and the scatter of data is similar. This is encouraging because experimental errors should be increased at the higher rates. The variation of results for BS 11 between the participating laboratories is shown in Table 6.

For the 50D material, high-rate tests show increasing toughness with increasing temperature, with a significant change apparent if the eight results between −50 and −20°C in Table 5 are grouped together and compared with the six tests between 20 and 50°C.

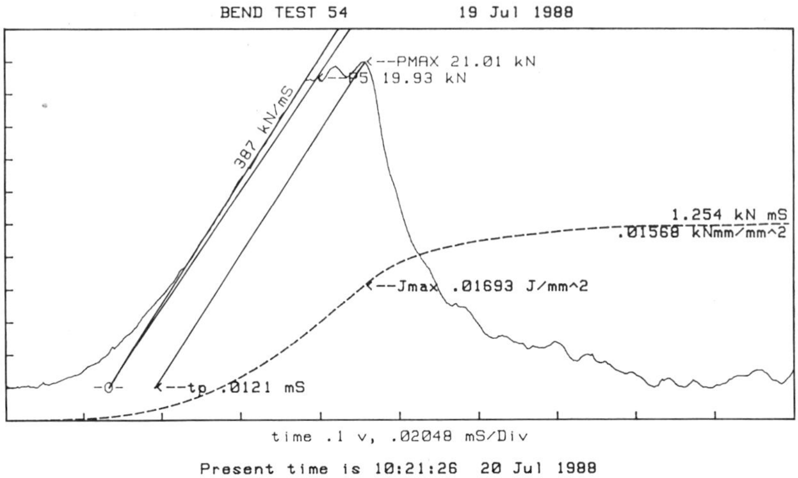


```
BS 5762              TEMP 20 DEG C
a= 20 mm    W= 40 mm    B= 20 mm
YS= 450 Nmm-2          NOM STRESS= 630.2 Nmm-2
COD= .0149 mm+ .0111 mm= .026 mm
Kq= 1678 Nmm-1.5    dK/dt= 32580 Nmm-1.5/mS
RATE= 5 mm/mS
```

FIG. 9—*Example of computerized record analysis for impact test on BS 11 bend specimen.*

CTOD Results

There are only a limited number of results from successful fully ductile dynamic tests. The two-bend specimen strain gage records that show a clear maximum load point give results that are in reasonable agreement. Two additional tests at Imperial College support these results. The strain gage also indicates pop-in behavior when it occurs, and the inflection point is clearly marked in many of the low-temperature 50D tests.

TABLE 5—*Fracture toughness results for the UKWP test program.*[a]

Specimen Type[a]	Number of Tests	Fracture Toughness, MN $m^{-1.5}$	Standard Deviation MN $m^{-1.5}$	Standard Deviation %	Valid B and a, mm
BS 11 bend	20	36.9	8.1	22	16.8
BS 11 Charpy	28	38.1	7.3	21	[17.9][b]
50D bend (−50°C)	1	52.3	...	...	17.5
50D bend (−30°C)	5	58.6	7.9	13	[22.0]
50D bend (−20°C)	2	60.8	7.3	12	[23.6]
50D bend (−50/−20°C)	8	58.4	6.9	12	[21.8]
50D bend (+20°C)	4	77.1	10.5	14	[38.0]
50D bend (+50°C)	2	78.7	6.7	8.5	[39.6]
50D bend (+20/+50°C)	6	77.6	8.6	11	[38.5]
50D Charpy (+20°C)	2	47.8	7.5	16	[14.6]

[a] BS 11 3PB specimens B and $a \cong 20$ mm, $\sigma_{y\,dyn}$ 450 MN m^{-2}. 50D 3PB specimens B and $a \cong 18$ mm, $\sigma_{y\,dyn}$ 625 MN m^{-2}.
[b] Invalid result using $2.5[K_{Ic}/\sigma_y]^2$ indicated thus [ ].

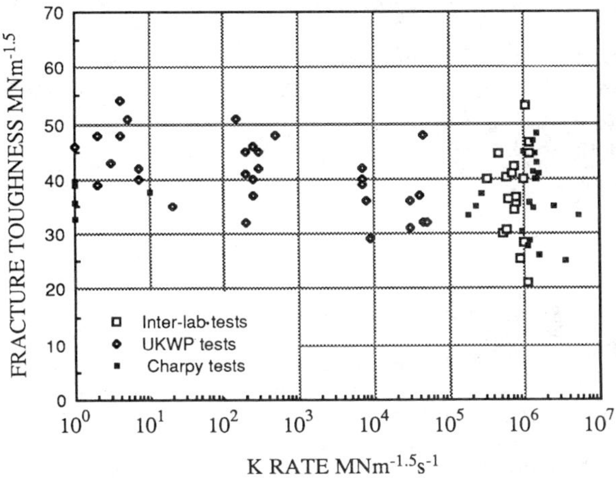

FIG. 10—*All BS 11 bend and Charpy fracture toughness results plus first UKWP static and higher rate tests as a function of rate.*

Conclusions

The testing and calibration technique developed on small Charpy specimens has been applied successfully to larger samples, using a range of servohydraulic, dropweight, and pendulum impact machines at K rates up to 10^6 MN $m^{-1.5}s^{-1}$. The performance of instruments and recording systems was compared, and the practical problems involved in successful dynamic fracture testing were assessed. Some problems in calibration and electrical interference were identified, and methods were developed to minimize their effect.

The proposed method can be used for assessing cleavage, ductile, and mixed-mode fracture using mainly existing standard techniques developed for BS 6729, and has been tested on two steels between −196 and +100°C. The fracture toughness results show good repeatability and no greater scatter than earlier tests at lower rates. Further confirmation may be required of the repeatability of fully ductile maximum load CTOD measurements, and a series of comparison tests between this method and the Dally and Sanford [*23*] procedure would be valuable. The next stage is to stimulate discussion of test results, encourage the widest use of the draft method, and incorporate it into a new British or European Standard.

TABLE 6—*Variation of mean fracture toughness results for BS 11 steel between participating laboratories in the UKWP test program.*

	BS 11 Charpy			BS 11 Bend		
Laboratory	K_{Ic}, MN $m^{-1.5}$	Standard Deviation	Number of Tests	K_{Ic}, MN $m^{-1.5}$	Standard Deviation	Number of Tests
Imperial College	41.6	3.3	16	36.8	7.2	6
Welding Institute	44.2	4.1	2	31.8	8.7	3
British Steel	28.2	3.7	3	48.0	3.6	3
Ruhr University	39.1	4.3	3	...	...	...
British Gas	25.9	0.3	2	36.2	0.5	3
CEGB-CERL	...	...	...	33.4	4.5	2
British Rail	29.6	0.8	2	33.5	4.5	3
ALL TESTS	38.1	7.3	28	36.9	8.1	20

Acknowledgments

The project was made possible by financial support from the Department of Trade and Industry and the six named U.K. organizations. The authors wish to thank Professor I. M. Allison, Dr. R. Hermann, Professor J. F. Kalthoff, Professor A. S. Kobayashi, Dr. C. C. Perry, members of the UKWP, and the technical staff of the collaborating laboratories for their valuable assistance.

References

[*1*] Christopher, P. R., in *Proceedings of the International Conference on Dynamic Fracture Toughness*, The Welding Institute, London, 1976.

[*2*] Christopher, P. R., *Advances in Fracture Research*, D. Francois, Ed., Pergamon Press, 1981, pp. 401–413.

[*3*] UK Briefing Group on Dynamic Fracture Testing, in *Welding Institute/ASM International Conference on Dynamic Fracture Toughness*, Vol. 2, 1981, pp. 1–11.

[*4*] Cannon, D. F. and MacGillivray, H. J., in *International Conference on Impact Loading and Dynamic Behaviour of Materials, Bremen*, Vol. 1, 1987, pp. 233–239.

[*5*] Baker, R. G. "Metrological Requirements in Fracture Mechanics Testing," Report for the Commission of the European Communities, Community Bureau of Reference, XII/353/88, 1988.

[*6*] Sumpter, J. D. G., Bird, J., Clarke, J. D., and Caudry, A. J., "Fracture Toughness of Ship Steels," RINA Spring Meeting, 1988.

[*7*] MacGillivray, H. J., Akum, E. R., Brookes-Johnson, P., and Grabulov, V., *BSSM/SEM International Conference on Advanced Measuring Techniques*, British Society for Strain Measurement/Society for Experimental Mechanics, London, 1987.

[*8*] Radon, J. C. and Turner, C. E., *Journal of Engineering Fracture Mechanics*, 1969, Vol 1, pp. 411–428.

[*9*] Server, W. L. and Tetelman, A. S., *Engineering Fracture Mechanics*, Vol. 4, 1972, p. 367.

[*10*] Kobayashi, A. S. "Dynamic Fracture Analysis by Dynamic Finite Element Method," in *Nonlinear and Dynamic Fracture Mechanics*, Perrone and Atluri, Eds., ASME AMD 35, 1979, pp. 19–36.

[*11*] Curr, R. M., MacGillivray, H. J., and Turner, C. E., "A Comparison of Some Elasto-Plastic Finite Element, Experimental Data and Simple Models for Notched Bars and Plates," in *IMechE Seminar on Dynamic Fracture Mechanics*, Institution of Mechanical Engineers, London, 1986.

[*12*] Ireland, D. R., *Welding Institute/ASM International Conference on Dynamic Fracture Toughness*, Vol. 1, 1977, pp. 47–62.

[*13*] Oi, K., *Experimental Mechanics*, Vol. 6, 1966, pp. 463–469.

[*14*] Brickle, L. W., *Experimental Mechanics*, Vol. 10, 1970, pp. 333–337.

[*15*] Loss, F. J., Hawthorn, J. R., and Griffis, C. A., "Fracture Toughness of Light Water Reactor Pressure Vessel Materials," Naval Research Laboratory Memorandum Report 3036, Washington, DC, April 1975.

[*16*] Giovanola, J. H. J., *Metals Handbook*, 9th ed. (American Society of Metals, Metals Park, OH, 1986, pp. 271–275.

[*17*] Mines, R. A. W. and Dutton, A. G. "An Analysis of the Hopkinson Pressure Bar-Loaded Instrumented Charpy Test Using Inertial Modelling Techniques," presented at the Seminar on Dynamic Fracture Mechanics, Imperial College, London, Oct. 1988.

[*18*] Mall, S., Kobayashi, A. S., and Loss, F. J., in *Crack Arrest Methodology and Applications, ASTM STP 711*, American Society for Testing and Materials, Philadelphia, 1980, pp. 70–85.

[*19*] Joyce, J. A. and Hackett, E. M., in *Fracture Mechanics: Seventeenth Volume, ASTM STP 905*, American Society for Testing and Materials, Philadelphia, 1986, pp. 741–774.

[*20*] Curry, D. A., Milne, I., and Gates, R. S., "The Influence of a High Loading Rate on the Fracture Behaviour of a Pressure Vessel Steel," CERL Report RD/L/2197N81, Central Electricity Research Laboratories, Leatherhead, England, 1982.

[*21*] Coque, H., Hudak, S. J., and Lindholm, U. S., *Journal de Physique*, C3, Suppl. 9, Vol. 3, 1988, pp. 347–352.

[*22*] Bruninghaus, K., Twickler, M., Twickler, R., and Dahl, W., *Journal de Physique*, C3, Suppl. 9, Vol. 3, 1988, pp. 301–306.

[*23*] Dally, J. W. and Sanford, R. J., *Experimental Mechanics*, Vol. 27, No. 4, 1987, pp. 381–388.

[*24*] Sharpe, W. N., Douglas, A. S., and Shapiro, J. M., "Dynamic Fracture Toughness Evaluation by Measurement of CTOD," Johns Hopkins University Report WNS/ASD-88-02, Baltimore, 1988.

[*25*] Kalthoff, J. F., *International Journal of Fracture*, Vol. 27, 1985, pp. 277–298.

[26] Hermann, R. and Holroyd, N., *Materials Science Technology*, Vol. 2, 1986, pp. 1238–1244.
[27] UKWP, "Project to Develop a Standard Method of Fracture Toughness Testing at Very High Loading Rates," Final Report DTI ref RTP2/155/78, Department of Trade and Industry, April 1990.
[28] Keegstra, P. N. R., Head, J. L., and Turner, C. E., "A Transient Finite Element Analysis of Unstable Crack Propagation in Some 2-Dimensional Geometries," in *Proceedings of the Fourth International Conference on Fracture, Waterloo*, Vol. 3, D. M. R. Taplin, Ed., University of Waterloo Press, Ontario, 1977, pp. 515–524.
[29] Brock, L. M., Jolles, M., and Schreodl, M., presented at the Joint ASME/ASCE Conference, Albuquerque, NM, 1985, pp. 1–4.